D1575420

A Primer of Quantum Chemistry

A PRIMER OF QUANTUM CHEMISTRY

F. C. GOODRICH

Professor of Chemistry
Clarkson College of Technology
Potsdam, New York

WILEY-INTERSCIENCE

a Division of John Wiley & Sons, Inc. New York · London · Sydney · Toronto

Library of Congress Cataloging in Publication Data

Goodrich, Frank Chauncey.
A primer of quantum chemistry.

Includes bibliographical references.
1. Quantum chemistry I. Title.

QD462.G66 541′.28 72-39307
ISBN 0-471-31490-0

Printed in the United States of America.

10 9 8 7 6 5 4 3 2 1

To

Chauncey Goodrich, 1836–1925

and

L. Carrington Goodrich, 1894–

of whom I am very proud

PREFACE

This book had its genesis in an informal series of seminars that I initiated at the Chevron Research Company in 1964, largely for the benefit of organic chemists who were repelled by the abstract formalism of most presentations of quantum chemistry but who nevertheless wanted to improve their understanding of the chemical bond. The course has gone through many evolutionary changes since its inclusion in 1965 as a one semester course in the curriculum at Clarkson College of Technology, but my bias for linear algebra as the most fundamental and least appreciated vehicle for presenting the subject has remained unmodified. There is consequently a strong emphasis on geometric concepts and intuitive images as a guide to the meaning of the formal algebraic or analytical procedures with a concommitant neglect of the fine details of, say, the solution of partial differential equations. The latter material is available in abundance in more advanced texts, to which the truly committed student will turn after completing the preliminary work of this course.

For their friendly support and interest, I am indebted to my former colleagues at the Chevron Research Company, particularly to Drs. R. C. Fox and M. J. R. Cantow. At Clarkson the material has been enriched by exposure to more than half a decade of seniors and first year graduate students. Finally I express my gratitude to several anonymous reviewers whose careful readings of the text in manuscript have saved it from a number of obscurities and outright errors.

F. C. GOODRICH

Potsdam, New York
November 1971

CONTENTS

A Primer of Quantum Chemistry

Chapter 1

FINITE DIMENSIONAL VECTOR SPACES

1.1 INTRODUCTION

The mathematical language of quantum mechanics is geometrical in nature. This statement may strike the beginning student as curious if he has ever peeped inside the cover of a textbook on the subject and been frightened by the complexity of the analytical formulas that disfigure almost every page. Essential to a sound grasp of the subject, however, is an ability to generalize one's intuitive perceptions of the geometry of two and three dimensions to spaces of a higher number of dimensions. The analytical tool necessary to do this is vector algebra.

1.2 VECTORS

A *scalar* is an ordinary number, real or complex. A *vector* $\mathbf{x}$ is a set of scalars written out in some order, either as a row or as a column:

$$\mathbf{x} = (x_1, x_2, \ldots, x_n)$$

The number n of *components* (entries) of the vector is said to be its *order* or *dimensionality*.

A geometric picture of a vector is conveniently obtained for $n = 2$ or $n = 3$ by interpreting the components of the vector as the coordinates of a point in n dimensional space. Then the directed line segment drawn from the origin of a Cartesian coordinate system to the point in question is a geometric realization of the vector (Figure 1.1).

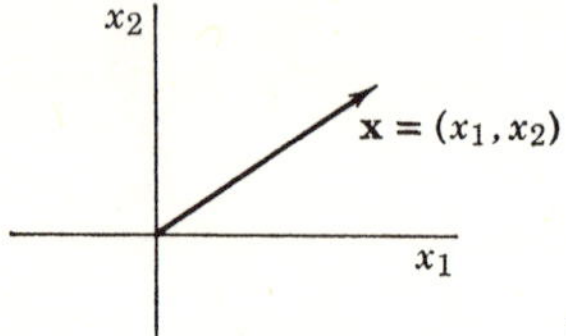

Figure 1.1 Geometric realization of a vector.

1.3 ALGEBRAIC PROPERTIES OF VECTORS

Two vectors are said to be equal only if they are of the same order and have all their components separately equal. Thus $\mathbf{x} = \mathbf{y}$ implies $x_1 = y_1$; $x_2 = y_2; \ldots ; x_n = y_n$, so that a single vector equality is equivalent to n scalar equalities.

Vector addition is defined by the equation

$$\begin{aligned}\mathbf{x} + \mathbf{y} &= (x_1, x_2, \ldots , x_n) + (y_1, y_2, \ldots , y_n) \\ &= (x_1 + y_1, x_2 + y_2, \ldots , x_n + y_n)\end{aligned}$$

It follows that vectors can be added only if they are of the same order and that vector addition is commutative: $\mathbf{x} + \mathbf{y} = \mathbf{y} + \mathbf{x}$. In two dimensions, vector addition is identical with the well known parallelogram rule of Galileo (Figure 1.2).

The *null* vector has all of its components identically zero: $\mathbf{0} = (0, 0, \ldots , 0)$, and we have the theorem $\mathbf{x} + \mathbf{0} = \mathbf{x}$.

The product of a vector by a scalar a is defined by

$$a\mathbf{x} = (ax_1, ax_2, \ldots , ax_n) = \mathbf{x}a$$

Illustrative of these definitions are the examples

(1) $(-1, 0, 3) + (1, -1, -4) = (0, -1, -1)$

(2) $2(0, -2, 2) = (0, -4, 4)$

(3) $4(1, -1, 0) - 2(1, -2, \frac{1}{2}) = (2, 0, -1)$

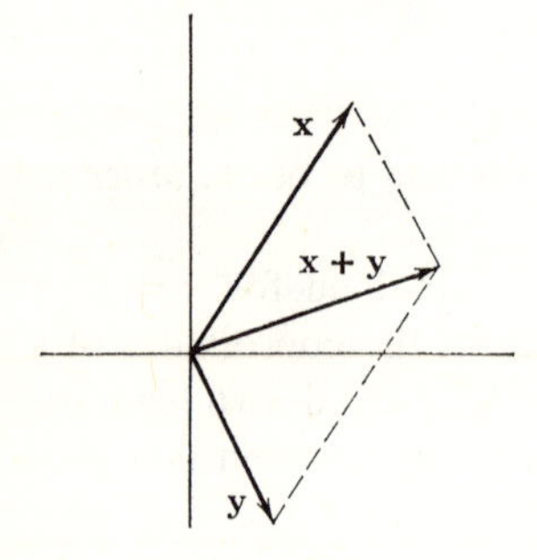

Figure 1.2 Vector addition: the parallelogram rule.

1.4 THE SCALAR PRODUCT

The *scalar product* of two vectors is not a vector but a scalar and is defined only for two vectors of the same order:

$$\mathbf{x} \cdot \mathbf{y} = x_1y_1 + x_2y_2 + \cdots + x_ny_n = \sum_1^n x_jy_j$$

Corresponding components of the two vectors are multiplied and the products summed. Note that $\mathbf{x} \cdot \mathbf{y} = \mathbf{y} \cdot \mathbf{x}$.

The scalar product of a vector by itself is the sum of the squares of its components. The positive square root of this sum is called the *norm* of the vector:

$$|x| = (\mathbf{x} \cdot \mathbf{x})^{1/2} = (x_1^2 + x_2^2 + \cdots + x_n^2)^{1/2} = \left[\sum_1^n x_j^2\right]^{1/2}$$

When interpreted geometrically in two and three dimensional space, the norm of a vector is by the Pythagorean theorem its length. It can also be demonstrated that the scalar product of two vectors in two or three dimensions is

$$\mathbf{x} \cdot \mathbf{y} = |x|\,|y| \cos \theta$$

where $0 \leq \theta \leq \pi$ is the angle between the two vectors. A vector of unit norm $|\mathbf{x}| = 1$ is said to be a *normalized* or *unit* vector.

If $\mathbf{x} \cdot \mathbf{y} = 0$, the vectors $\mathbf{x}$ and $\mathbf{y}$ are said to be *orthogonal*. From the formula $\mathbf{x} \cdot \mathbf{y} = |x|\,|y| \cos \theta$, we observe that this happens in two and three dimensions when $\theta = \pi/2$ radians $= 90°$, that is, when the vectors are perpendicular. Note that formally $\mathbf{x} \cdot \mathbf{0} = 0$ for every $\mathbf{x}$, so that the null vector is orthogonal to every other vector.

To illustrate these operations, consider in Figure 1.2

$$\mathbf{x} = (2, 3); \qquad \mathbf{y} = (1, -2)$$

The scalar product is $\mathbf{x} \cdot \mathbf{y} = (2)(1) + (3)(-2) = -4$. The norms are

$$|x| = (2^2 + 3^2)^{1/2} = 3.606$$
$$|y| = (1^2 + (-2)^2)^{1/2} = 2.236$$

The cosine of the angle between $\mathbf{x}$ and $\mathbf{y}$ is therefore

$$\cos \theta = \frac{\mathbf{x} \cdot \mathbf{y}}{|x|\,|y|} = \frac{-4}{(3.606)(2.236)} = -0.496$$

or $\theta = 119°45'$.

PROBLEMS

1.1 Calculate to the nearest minute the angle between the following pairs of vectors:

(1) $(1, 0, -1)$
$(2, 1, \quad 0)$

(2) $(0, -1, 0, 3)$
$(3, \quad 1, 0, 0)$

1.2 Normalize the vector $\mathbf{x} = (1, -2, 0)$ by dividing it through by its own norm, $\mathbf{E} = \mathbf{x}/|x|$. Note that $\mathbf{E}$ is a unit vector.

1.3 A unit cell of the sodium chloride crystal has the shape of a tiny cube in which closed circles represent sodium ions and open circles represent chloride ions. At the center of this cube (not shown) is a sodium ion.

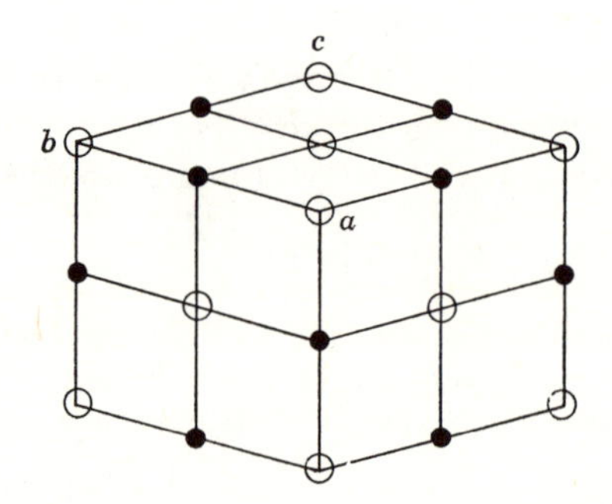

Calculate:

(1) The angle between vectors drawn from the central sodium ion to chloride ions a and b.

(2) The angle between vectors drawn from the central ion to chlorides a and c.

1.5 MATRICES

In this book a *matrix* will always be a square array of n^2 scalars

$$A = \begin{pmatrix} a_{11} & a_{12} & \cdots & a_{1n} \\ a_{21} & a_{22} & \cdots & a_{2n} \\ \cdot & \cdot & & \cdot \\ \cdot & \cdot & & \cdot \\ \cdot & \cdot & & \cdot \\ a_{n1} & a_{n2} & \cdots & a_{nn} \end{pmatrix}$$

arranged in some order, where the number n of rows or columns is the *order* or *dimensionality* of the matrix. Two matrices are said to be equal only if they are of the same order and have corresponding *components* (entries) identical. Thus $A = B$ implies $a_{ij} = b_{ij}$ so that a single matrix equality is equivalent to n^2 scalar equalities.

Matrices exist in which the number of rows is not equal to the number of columns, but such rectangular arrays will not be needed in this book.

1.6 MATRIX ALGEBRA

Matrices of the same order may be added to yield another matrix under the rule $A + B = C$ implies $c_{ij} = a_{ij} + b_{ij}$, that is, corresponding components of A and B are added to yield C. The *null* matrix $\mathcal{O}$ has all of its components equal to zero, and for this matrix $A + \mathcal{O} = A$.

For vectors we defined a scalar product such that the result of the multiplication is not a vector but a scalar. For matrices, however, we define a matrix product such that two matrices of the same order may be multiplied together to yield a third matrix. If $AB = C$, then the arithmetic rule whereby the components of C are calculated from the components of A and B is

$$c_{ij} = \sum_{k=1}^{n} a_{ik} b_{kj}$$

Note that c_{ij}, which is the component of C located at the intersection of the ith row with the jth column, is calculated from the components of the ith row of A and the components of the jth column of B (Figure 1.3).

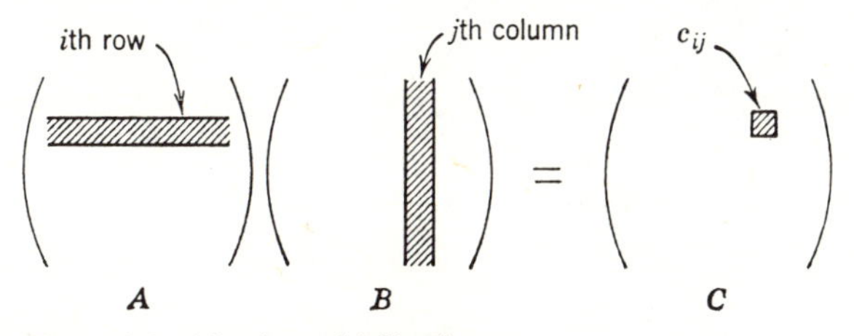

Figure 1.3 Matrix multiplication.

The student unfamiliar with the arithmetic details of matrix multiplication will do well to work out a few examples. Thus if

$$A = \begin{pmatrix} -1 & 0 & 1 \\ 2 & 3 & 0 \\ 0 & -2 & 1 \end{pmatrix}$$

$$B = \begin{pmatrix} 7 & 0 & 2 \\ -6 & 3 & 1 \\ 4 & 0 & -3 \end{pmatrix}$$

then c_{11} is calculated from the first row of A and the first column of B:

$$c_{11} = (-1)(7) + (0)(-6) + (1)(4) = -3$$

Similarly

$$c_{12} = (-1)(0) + (0)(3) + (1)(0) = 0$$

$$c_{13} = (-1)(2) + (0)(1) + (1)(-3) = -5$$

The student should continue this process to find

$$C = \begin{pmatrix} -3 & 0 & -5 \\ -4 & 9 & 7 \\ 16 & -6 & -5 \end{pmatrix}$$

It cannot escape him that the numerical procedure in calculating a matrix product is identical in vector algebra with that of forming n^2 scalar products.

PROBLEM

1.4 Given the matrices

$$A = \begin{pmatrix} 1 & 1 \\ -1 & -1 \end{pmatrix}, \qquad B = \begin{pmatrix} 1 & 0 \\ -1 & 0 \end{pmatrix}$$

calculate the products AB and BA.

1.7 MULTIPLICATION OF A MATRIX TIMES A VECTOR

The product of a matrix times a vector is another vector:

$$A\mathbf{x} = \mathbf{y}$$

where $y_i = \sum_{j=1}^{n} a_{ij}x_j$. It is conventional in carrying out this operation to write the matrix on the left and the vector as a column on the right:

$$\begin{pmatrix} a_{11} & \cdots & a_{1n} \\ \cdot & & \cdot \\ \cdot & & \cdot \\ \cdot & & \cdot \\ a_{n1} & \cdots & a_{nn} \end{pmatrix} \begin{pmatrix} x_1 \\ \cdot \\ \cdot \\ \cdot \\ x_n \end{pmatrix} = \begin{pmatrix} y_1 \\ \cdot \\ \cdot \\ \cdot \\ y_n \end{pmatrix}$$

Then the arithmetic follows the same row-column pattern as matrix multiplication in that y_i is calculated from the elements of the ith row of A and those of the single column of $\mathbf{x}$. Note that a matrix enters into mathematics as an operator which sends a vector into another vector, changing its direction as well as its length. We shall see that the mechanical state of a system in quantum mechanics is described by a vector. Only those vectors are acceptable as state vectors that transform in a definite way when operated on by certain matrices.

An important theorem concerning the multiplication of a matrix times a vector is that the operation is distributive:

$$A(\mathbf{x} + \mathbf{y}) = A\mathbf{x} + A\mathbf{y}$$

1.8 LINEAR TRANSFORMATIONS AND MATRICES

The totality of all vectors of order n is known as an n dimensional vector space. Because the result of operating on a vector with a matrix is another vector, the result of operating on a vector space with a matrix is to map the space onto itself. Such a mapping is known as a *linear transformation.* Let us consider some examples.

Suppose that

$$D = \begin{pmatrix} 2 & 0 \\ 0 & 2 \end{pmatrix}$$

and consider the effect that D has on four unit vectors lying along the coordinate axes of a two dimensional space. That is, we form the products

$$D\mathbf{x}_1 = \begin{pmatrix} 2 & 0 \\ 0 & 2 \end{pmatrix}\begin{pmatrix} 1 \\ 0 \end{pmatrix} = \begin{pmatrix} 2 \\ 0 \end{pmatrix}; \qquad D\mathbf{x}_2 = \begin{pmatrix} 2 & 0 \\ 0 & 2 \end{pmatrix}\begin{pmatrix} 0 \\ 1 \end{pmatrix} = \begin{pmatrix} 0 \\ 2 \end{pmatrix}$$

$$D\mathbf{x}_3 = \begin{pmatrix} 2 & 0 \\ 0 & 2 \end{pmatrix}\begin{pmatrix} -1 \\ 0 \end{pmatrix} = \begin{pmatrix} -2 \\ 0 \end{pmatrix}; \qquad D\mathbf{x}_4 = \begin{pmatrix} 2 & 0 \\ 0 & 2 \end{pmatrix}\begin{pmatrix} 0 \\ -1 \end{pmatrix} = \begin{pmatrix} 0 \\ -2 \end{pmatrix}$$

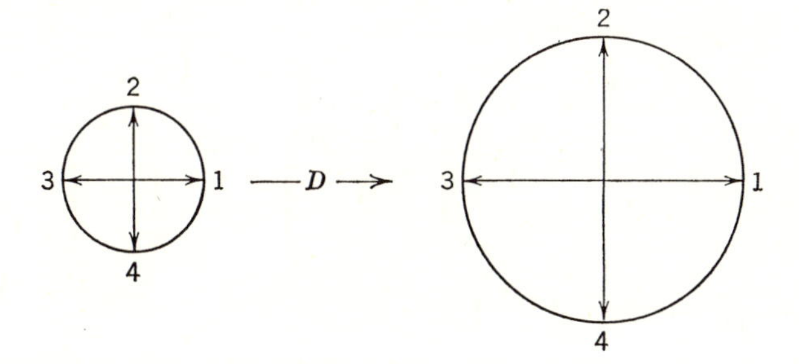

Figure 1.4 Linear transformation with a dilation matrix.

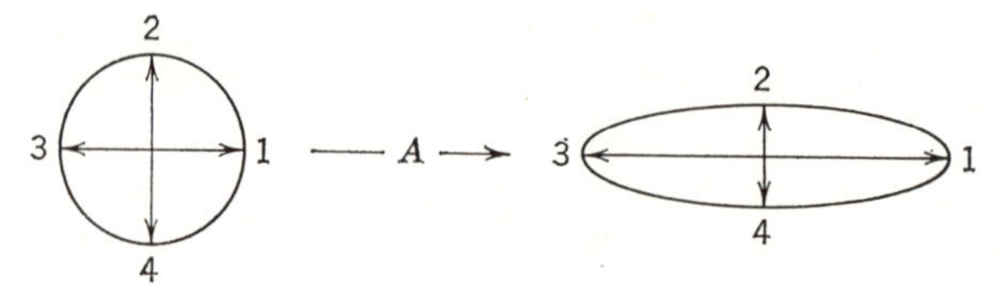

Figure 1.5 Linear transformation with a strain matrix.

In each case the effect of D has been to leave the vector unchanged in direction but doubled in length, and the student may readily satisfy himself that this holds true for every vector in the space, so that, for instance, the unit circle is sent into a circle of radius 2. Geometrically the picture is Figure 1.4, and we may reasonably describe D by saying that it is a dilation matrix, for it expands ("blows up") the space uniformly in every direction.

Now consider the matrix

$$A = \begin{pmatrix} 2 & 0 \\ 0 & \frac{1}{2} \end{pmatrix}$$

and let A operate on our four unit vectors $\mathbf{x}_1$, $\mathbf{x}_2$, $\mathbf{x}_3$, $\mathbf{x}_4$ and more generally on all unit vectors (whose termini must therefore lie on the unit circle). The student may verify that the resulting geometric picture is Figure 1.5, so that vectors along the horizontal axis are doubled in length, those along the vertical axis are halved in length, and all other vectors are changed in both direction and length in such a way that circles concentric about the origin are sent into ellipses. Because the space is stretched in one direction and shrunken in another, A is an example of a strain matrix.

As another example, take

$$R = \begin{pmatrix} 1/\sqrt{2} & -1/\sqrt{2} \\ 1/\sqrt{2} & 1/\sqrt{2} \end{pmatrix}$$

and let it operate on our unit vectors. There results Figure 1.6, so that linear transformation with R rotates the space 45° counterclockwise.

Finally an experiment with the matrix

$$P = \begin{pmatrix} 1 & 0 \\ 0 & 0 \end{pmatrix}$$

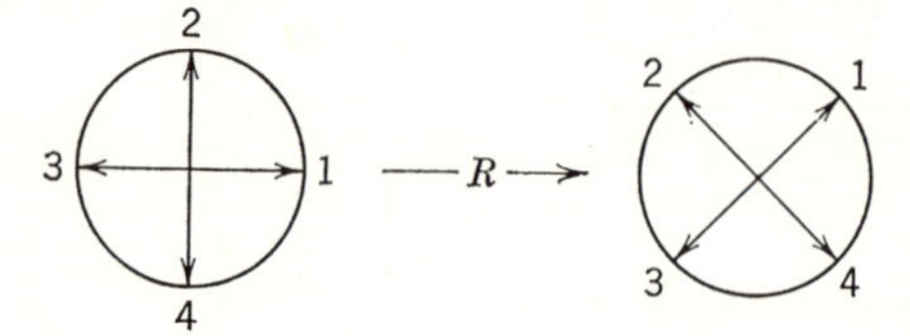

Figure 1.6 Linear transformation with a rotation matrix.

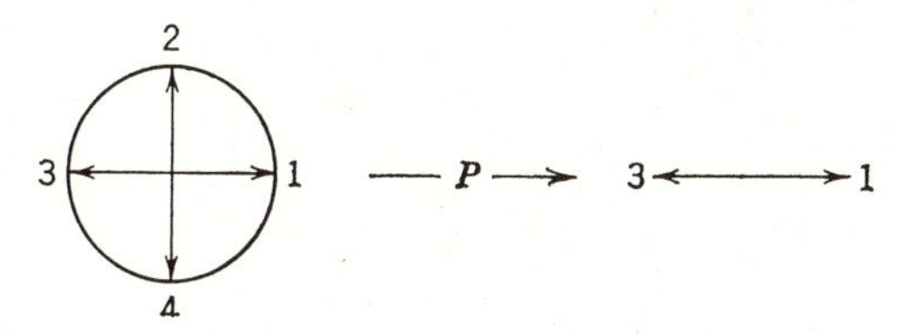

Figure 1.7 Linear transformation with a projection matrix.

will show that while $\mathbf{x}_1$ and $\mathbf{x}_3$ are reproduced identically, $\mathbf{x}_2$ and $\mathbf{x}_4$ are annihilated (sent into the null vector) as sketched in Figure 1.7. P evidently projects the entire space onto the horizontal axis.

1.9 SUCCESSIVE LINEAR TRANSFORMATIONS AND MATRIX MULTIPLICATION

Once a vector space has been transformed by a linear transformation, it may evidently be transformed again by another (or the same) linear transformation. This if $\mathbf{y} = A\mathbf{x}$, then $A\mathbf{y} = A(A\mathbf{x}) = A^2\mathbf{x}$. Geometrically this looks like Figure 1.8, in which

$$A^2 = \begin{pmatrix} 2 & 0 \\ 0 & \frac{1}{2} \end{pmatrix} \begin{pmatrix} 2 & 0 \\ 0 & \frac{1}{2} \end{pmatrix} = \begin{pmatrix} 4 & 0 \\ 0 & \frac{1}{4} \end{pmatrix}$$

The row-column rule for matrix multiplication was invented in order to reproduce algebraically these geometric transformations, so that the result of two linear transformations applied successively is another linear transformation. Another way of stating the same fact is that the product of two matrices is another matrix.

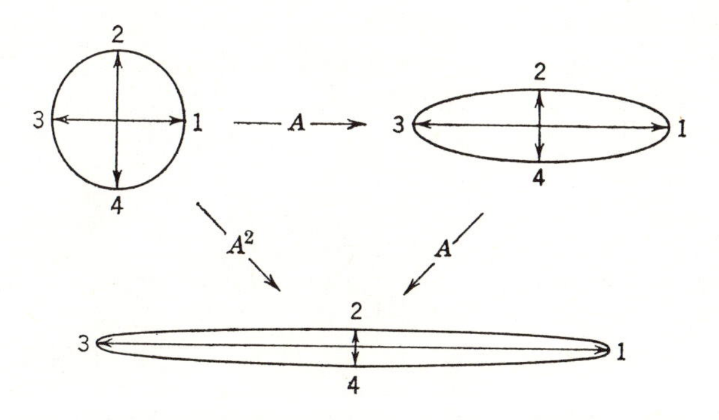

Figure 1.8 Successive linear transformations.

Similarly

$$R^2 = \begin{pmatrix} 1/\sqrt{2} & -1/\sqrt{2} \\ 1/\sqrt{2} & 1/\sqrt{2} \end{pmatrix} \begin{pmatrix} 1/\sqrt{2} & -1/\sqrt{2} \\ 1/\sqrt{2} & 1/\sqrt{2} \end{pmatrix} = \begin{pmatrix} 0 & -1 \\ 1 & 0 \end{pmatrix}$$

and the student may verify by experiment that this states that two counterclockwise rotations of 45° yield a single counterclockwise rotation of 90°. Furthermore, $P^2 = P$, and the student is invited to interpret geometrically this algebraic property of P.

1.10 THE COMMUTATIVITY OF MATRIX MULTIPLICATION

The algebraic rule $AB = BA$ is in general not obeyed by matrices, which is to say that matrix multiplication is not commutative. The student will find by actual calculation on the matrices of Section 1.8 that $AR \neq RA$, and geometrically this states that the order of application of linear transformations makes a difference in the final result. In the example chosen, if we first stretch the space along the horizontal axis and then rotate it 45° counterclockwise, we obtain a different result than if we first rotated and then stretched along the horizontal axis (Figure 1.9).

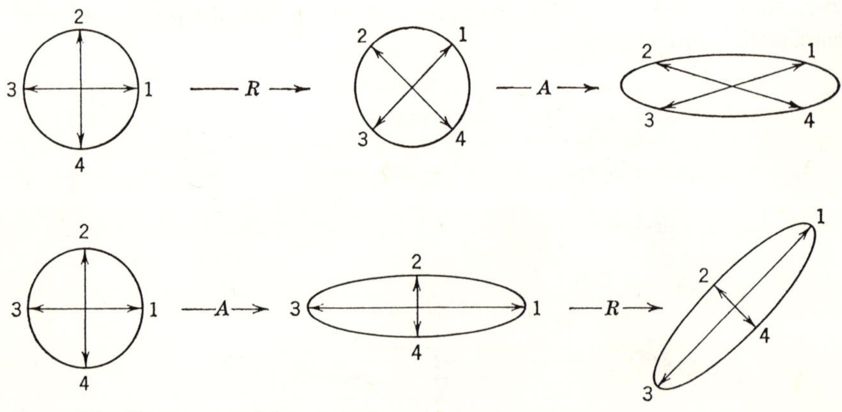

Figure 1.9 Noncommutativity of matrix multiplication, $AR \neq RA$.

PROBLEMS

1.5 An arbitrary unit vector in two dimensions can be written $\mathbf{x} = (\cos\alpha, \sin\alpha)$. For

$$K = \begin{pmatrix} \frac{5}{3} & \frac{4}{3} \\ \frac{4}{3} & \frac{5}{3} \end{pmatrix} \quad \text{and} \quad R = \begin{pmatrix} 1/\sqrt{2} & -1/\sqrt{2} \\ 1/\sqrt{2} & 1/\sqrt{2} \end{pmatrix}$$

plot on a sheet of graph paper the termini of $K\mathbf{x}$ and on another sheet the termini of $R\mathbf{x}$ for the eight unit vectors corresponding to $\alpha = m\pi/4$; $m = 0, 1, 2, \ldots, 7$. Label each terminus with the value of m used.

1.6 Calculate $(RK)\mathbf{x}$ and $(KR)\mathbf{x}$ and plot on two additional sheets of graph paper the termini of the same eight vectors appropriately labeled.

1.11 THE UNIT MATRIX, RECIPROCAL MATRICES

The matrix

$$\begin{pmatrix} 1 & 0 & 0 & \cdots & 0 \\ 0 & 1 & 0 & \cdots & 0 \\ \cdot & & & & \cdot \\ \cdot & & & & \cdot \\ \cdot & & & & \cdot \\ 0 & 0 & \cdots & \cdots & 1 \end{pmatrix}$$

which has unity along the *main diagonal* and zeros elsewhere is known as the *unit* or identity matrix. It is evidently a linear transformation that maps each vector in a vector space into itself, leaving the space unaltered. The student should check for himself the truth of the following theorems: $I\mathbf{x} = \mathbf{x}$ for every vector $\mathbf{x}$, and $IA = AI = A$ for every matrix A.

If two matrices H and G bear the relationship $GH = I$, they are said to be *reciprocal* to each other and they are written $H = G^{-1}$ or $G = H^{-1}$. Alternatively G^{-1} is said to be the *inverse* of G. An example from Section 1.8 is

$$AA^{-1} = \begin{pmatrix} 2 & 0 \\ 0 & \frac{1}{2} \end{pmatrix}\begin{pmatrix} \frac{1}{2} & 0 \\ 0 & 2 \end{pmatrix} = \begin{pmatrix} 1 & 0 \\ 0 & 1 \end{pmatrix} = I$$

The linear transformation A^{-1} evidently undoes what A does—A^{-1} shrinks along the horizontal axis and stretches along the vertical axis to just a sufficient degree to cancel the effect of A. Similarly

$$RR^{-1} = \begin{pmatrix} 1/\sqrt{2} & -1/\sqrt{2} \\ 1/\sqrt{2} & 1/\sqrt{2} \end{pmatrix} \begin{pmatrix} 1/\sqrt{2} & 1/\sqrt{2} \\ -1/\sqrt{2} & 1/\sqrt{2} \end{pmatrix} = \begin{pmatrix} 1 & 0 \\ 0 & 1 \end{pmatrix} = I$$

Not every matrix, however, possesses an inverse. The matrix P is an example of this, and in general a linear transformation which involves a projection cannot be inverted. The geometric interpretation of this is that a *singular* linear transformation (one that does *not* possess an inverse) maps many vectors in the original space into some subspace. This process cannot be inverted uniquely. On the other hand, nonsingular linear transformations (those matrices that possess an inverse) map the vector space onto itself in a one to one fashion, and this process can be inverted.

Some theorems concerning reciprocal matrices are the following:

1. $AA^{-1} = A^{-1}A = I.$
2. $(A^{-1})^{-1} = A.$
3. The inverse of a matrix, if it exists, is unique.

I have not given any method of computing the inverse of a matrix, for we shall not require it in our work. The problem is identical with finding a general solution to a set of n linear equations in n unknowns, and many methods are known for computing this solution[1,2] (Cramer's rule, for example). A convenient test for the existence of an inverse is to examine the determinant of the matrix formed by treating the rows and columns of the matrix as the components of a determinant. If this determinant is zero, the matrix is singular. If it is not zero, the matrix is nonsingular.

PROBLEMS

1.7 Multiply

$$\begin{pmatrix} 0 & -1 & 3 \\ 2 & 0 & 4 \\ -1 & 0 & 1 \end{pmatrix} \begin{pmatrix} 0 & \frac{1}{6} & -\frac{2}{3} \\ -1 & \frac{1}{2} & 1 \\ 0 & \frac{1}{6} & \frac{1}{3} \end{pmatrix}$$

1.8 The general rotation matrix in two dimensions is

$$R(\theta) = \begin{pmatrix} \cos\theta & \sin\theta \\ -\sin\theta & \cos\theta \end{pmatrix}$$

in which θ is the angle of rotation. Prove that

$$R(\theta)R(\omega) = R(\theta + \omega)$$

1.12 THE THEORY OF LINEAR DEPENDENCE

The problem frequently arises in practice of expressing a given vector $\mathbf{y}$ as a *linear combination* of n given vectors $\mathbf{x}_j$:

$$\mathbf{y} = c_1\mathbf{x}_1 + c_2\mathbf{x}_2 + \cdots + c_n\mathbf{x}_n = \sum_1^n c_j\mathbf{x}_j$$

where the c_j are scalar constants to be determined. This single vector equation is equivalent to n ordinary scalar equations, which, when written out in detail, take the form

$$\begin{aligned} y_1 &= c_1x_{11} + c_2x_{12} + \cdots + c_nx_{1n} \\ y_2 &= c_1x_{21} + c_2x_{22} + \cdots + c_nx_{2n} \\ &\cdot \\ &\cdot \\ &\cdot \\ y_n &= c_1x_{n1} + c_2x_{n2} + \cdots + c_nx_n \end{aligned}$$

in which

$$\begin{aligned} \mathbf{x}_1 &= (x_{11}, x_{21}, \ldots, x_{n1}) \\ \mathbf{x}_2 &= (x_{12}, x_{22}, \ldots, x_{n2}) \\ &\cdots \end{aligned}$$

Our problem is: given the x_{ij} and the y_i, find the c_j. Stated in this way we see that we are asked to find the solution to a set of n linear equations in n unknowns. Such a set of equations has a unique solution only if the determinant of coefficients $|x_{ij}|$ does not vanish, and in general a set of n vectors in n dimensional space is said to be *linearly independent* if their components when assembled into the rows or columns of a determinant do not yield a zero determinant. Otherwise the vectors are said to be *linearly dependent*.

The geometric meaning of linear dependence is illustrated by an example: test the linear dependence of the vector sets

$$\text{(a)}\quad \mathbf{x}_1 = (1, -2) \qquad \text{(b)}\quad \mathbf{x}_1 = (1, -3)$$
$$\mathbf{x}_2 = (-2, 4) \qquad \mathbf{x}_2 = (1, 2)$$

We construct the determinants

$$\text{(a)}\quad \begin{vmatrix} 1 & -2 \\ -2 & 4 \end{vmatrix} = 0 \qquad \text{(b)}\quad \begin{vmatrix} 1 & -3 \\ 1 & 2 \end{vmatrix} = 5$$

whence it follows that (a) is linearly dependent and (b) is linearly independent. The student will find it suggestive to draw diagrams of each vector pair (a) and (b) after the manner of Figures 1.1 and 1.2. The vector set (a) will be observed to be collinear while the vectors of set (b) meet at an obtuse angle.

Generally speaking, linearly dependent sets of vectors in two dimensions are collinear; linearly dependent sets of vectors in three dimensions are coplanar; and so on. A linearly independent set of vectors completely spans a vector space without being localized to some subspace of a lower number of dimensions.

1.13 BASES AND COORDINATE SYSTEMS

We have just seen that for a set of n linearly independent $\mathbf{x}_j$, a unique expansion exists for any arbitrary vector $\mathbf{y}$:

$$\mathbf{y} = c_1\mathbf{x}_1 + \cdots + c_n\mathbf{x}_n$$

If the $\mathbf{x}_j$ are linearly dependent, the expansion either does not exist or is not unique. A set of n linearly independent vectors may thus be used as a *basis* or a *coordinate system* in a space of n dimensions in the sense that any arbitrary vector in the space may be expressed as a linear combination of the basis vectors in one and only one way. This is frequently a practical thing to do, and we have the theorem that if the $\mathbf{x}_j$ are linearly independent, then

$$c_1\mathbf{x}_1 + c_2\mathbf{x}_2 + \cdots + c_n\mathbf{x}_n = b_1\mathbf{x}_1 + b_2\mathbf{x}_2 + \cdots + b_n\mathbf{x}_n$$

implies $c_1 = b_1$; $c_2 = b_2$; . . . ; $c_n = b_n$.

1.14 ORTHONORMAL BASES

From a purely mathematical point of view, no one basis has any preference over any other, and relations between vectors can be expressed just as well in all coordinate systems. From the point of view of the investigator interested in results, however, *orthonormal* bases have a distinct advantage, for the basis vectors can be manipulated very simply. An orthonormal set of vectors $\mathbf{E}_1, \mathbf{E}_2, \ldots, \mathbf{E}_n$ has the property

$$\begin{aligned} \mathbf{E}_i \cdot \mathbf{E}_j = \delta_{ij} &= 1 \qquad \text{if } i = j \\ &= 0 \qquad \text{otherwise} \end{aligned}$$

The symbol δ_{ij} is called the Kronecker delta. An example of an orthonormal coordinate system is

$$\begin{aligned} \mathbf{e}_1 &= (1, 0, 0, \ldots, 0) \\ \mathbf{e}_2 &= (0, 1, 0, \ldots, 0) \\ &\;\;\vdots \\ \mathbf{e}_n &= (0, 0, 0, \ldots, 1) \end{aligned}$$

These are the usual Cartesian axes in which we conventionally express vectors. The set $\mathbf{e}_j$ is linearly independent

$$|e_{ij}| = |\delta_{ij}| = 1 \neq 0$$

and orthonormal $\mathbf{e}_i \cdot \mathbf{e}_j = \delta_{ij}$. Other orthonormal bases exist, however, and for example in two dimensions the set

$$\begin{aligned} \mathbf{E}_1 &= (1/\sqrt{2}, \quad 1/\sqrt{2}) \\ \mathbf{E}_2 &= (1/\sqrt{2}, \quad -1/\sqrt{2}) \end{aligned} \tag{1.1}$$

is linearly independent and orthonormal:

$$\mathbf{E}_1 \cdot \mathbf{E}_1 = \mathbf{E}_2 \cdot \mathbf{E}_2 = 1$$

$$\mathbf{E}_1 \cdot \mathbf{E}_2 = \mathbf{E}_2 \cdot \mathbf{E}_1 = 0$$

1.15 EXPANSIONS IN ORTHONORMAL BASES

The determination of the unknown coefficients c_j in the expansion of a known vector $\mathbf{y}$ in terms of a given orthonormal basis is particularly simple:

$$\mathbf{y} = c_1\mathbf{E}_1 + c_2\mathbf{E}_2 + \cdots + c_n\mathbf{E}_n$$

If the $\mathbf{E}_j$ were not orthonormal, the problem, as we have seen, would be that of the solution of n linear equations in n unknowns. For the $\mathbf{E}_j$ orthonormal, however, take the scalar product of each side of the above equation by $\mathbf{E}_1$. The result is

$$\mathbf{E}_1 \cdot \mathbf{y} = c_1\mathbf{E}_1 \cdot \mathbf{E}_1 + 0 + 0 + \cdots + 0 = c_1$$

The method is quite general, and we may write $c_j = \mathbf{E}_j \cdot \mathbf{y}$.

In two dimensions a calculation using the orthonormal basis (1.1) is typical. Let $\mathbf{y} = (1, -2)$ and find c_1, c_2 for which

$$\mathbf{y} = c_1\mathbf{E}_1 + c_2\mathbf{E}_2$$

We have $c_1 = \mathbf{E}_1 \cdot \mathbf{y} = -1/\sqrt{2}$; $c_2 = \mathbf{E}_2 \cdot \mathbf{y} = 3/\sqrt{2}$, so that

$$\mathbf{y} = (-1/\sqrt{2})\mathbf{E}_1 + (3/\sqrt{2})\mathbf{E}_2$$

If the student will substitute from (1.1) into this formula he will find that the vector $\mathbf{y}$ is reproduced identically.

PROBLEM

1.9 Given the orthonormal set of vectors

$$\mathbf{E}_1 = (1/\sqrt{5}, 0, -2/\sqrt{5})$$

$$\mathbf{E}_2 = (0, 1, 0)$$

$$\mathbf{E}_3 = (2/\sqrt{5}, 0, 1/\sqrt{5})$$

find the coefficients c_j in the expansion $\mathbf{y} = c_1\mathbf{E}_1 + c_2\mathbf{E}_2 + c_3\mathbf{E}_3$ for $\mathbf{y} = (2, -1, 1)$.

1.16 TRANSFORMATION THEORY

Imagine that we have some physical property, such as a velocity or force, which can be represented by a vector in ordinary three dimensional space. It is apparent that depending on how we draw our set of orthogonal axes (or even if we use nonorthogonal axes) the mathematical representation of the physical vector (but not the vector itself) will change. Such changes do not, of course, have any effect on the physical property represented by the vector but only reflect the various mathematical choices open to the investigator. Because the mathematician has this freedom of choice in the way he can describe a given physical situation, he quite properly makes choices in such a way as to simplify the algebraic or computational details of his work. In other words, he often needs a method of calculating the components of a vector in a new coordinate system when he is given its components in some other coordinate system. This calculation can be effected by matrix algebra, and it is shown in texts[1,2] upon the subject that if the components of a vector in the old coordinate system are $\mathbf{x}$ and the components of the same (physical) vector in the new coordinate system are $\mathbf{x}'$, then the numbers $\mathbf{x}'$ are related to the numbers $\mathbf{x}$ by $\mathbf{x}' = M\mathbf{x}$ where M is the *matrix of the transformation.* Of particular importance in practical work is the *orthogonal transformation* by means of which one set of orthogonal axes is converted into another. Intuitively we should expect an orthogonal transformation in two or three dimensional space to correspond to a rotation or a reflection of coordinate axes, and it is the generalization of this expectation to spaces of a higher number of dimensions which has led to the n dimensional *orthogonal matrix.* If a matrix M is orthogonal, it has the remarkable property that its n rows and its n columns each separately constitute a set of n orthonormal vectors.

We shall consider further properties of these matrices in due course; for now it is sufficient to accept the idea that if $\mathbf{E}_1, \mathbf{E}_2, \ldots, \mathbf{E}_n$ are orthonormal, then for M an orthogonal matrix, the vectors $\boldsymbol{\epsilon}_1 = M\mathbf{E}_1, \boldsymbol{\epsilon}_2 = M\mathbf{E}_2, \ldots, \boldsymbol{\epsilon}_n = M\mathbf{E}_n$ also constitute an orthonormal set.

Examples of orthogonal matrices are I and the rotation matrix R of Section 1.8.

PROBLEM

1.10 (1) Through what angle does the matrix $K = \begin{pmatrix} 1 & 2 \\ 2 & 1 \end{pmatrix}$ rotate the vector $\mathbf{v} = (1, 0)$? (I.e., find the angle between $\mathbf{v}$ and $K\mathbf{v}$.) By what factor is $\mathbf{v}$ lengthened by transformation with K?

(2) Repeat these calculations for $\mathbf{u} = (1, 1)$.

1.11 The general rotation matrix in two dimensions may be written as

$$R = \begin{pmatrix} \cos\varphi & \sin\varphi \\ -\sin\varphi & \cos\varphi \end{pmatrix}$$

Prove that the scalar product of two vectors is invariant under a linear transformation with R. That is, for $\mathbf{x} = (x_1, x_2)$ and $\mathbf{y} = (y_1, y_2)$ show that $\mathbf{x} \cdot \mathbf{y}$ is the same as $(R\mathbf{x}) \cdot (R\mathbf{y})$. What special geometric interpretation does this have if $\mathbf{x} = \mathbf{y}$?

1.17 THE EIGENVALUE-EIGENVECTOR PROBLEM

If a vector $\mathbf{x}$ has the property with respect to a given matrix K of satisfying

$$K\mathbf{x} = \lambda\mathbf{x}$$

where λ is a scalar, $\mathbf{x}$ is said to be an *eigenvector* of K with *eigenvalue* λ. Geometrically, if a vector space undergoes a linear transformation, those vectors that are altered in length but not in direction are eigenvectors of that linear transformation. This statement must be interpreted to include "negative" length—that is, the vector may be reflected backward through the origin and still be an eigenvector, in this case with a negative eigenvalue. Some examples:

1. Given $A = \begin{pmatrix} 2 & 0 \\ 0 & \frac{1}{2} \end{pmatrix}$. Then $A\mathbf{x} = \lambda\mathbf{x}$ is satisfied if $\mathbf{x} = \begin{pmatrix} 1 \\ 0 \end{pmatrix}$ and $\lambda = 2$, or if $\mathbf{x} = \begin{pmatrix} 0 \\ 1 \end{pmatrix}$ and $\lambda = \frac{1}{2}$.

2. The rotation matrix R of Section 1.8 possesses no nonnull eigenvectors or eigenvalues of the type we have been studying, that is, whose components belong to the field of real numbers.

3. The projection matrix P of Section 1.8 possesses eigenvectors of the form $\mathbf{x} = \begin{pmatrix} x \\ 0 \end{pmatrix}$ with arbitrary x, all of which have eigenvalue $\lambda = 1$.

4. Every vector in n dimensional space is an eigenvector of the identity matrix I, all with eigenvalue $\lambda = 1$.

1.18 THE SECULAR EQUATION

In quantum mechanics, the eigenvalue-eigenvector problem presents itself in the following way: given a matrix K, find its eigenvalues and eigenvectors. In matrix terminology we are to find all λ's and $\mathbf{x}$'s that satisfy

$K\mathbf{x} = \lambda\mathbf{x}$. Written out in full, this vector equation is equivalent to n scalar equations:

$$\begin{aligned} k_{11}x_1 + k_{12}x_2 \quad \cdots \quad + k_{1n}x_n &= \lambda x_1 \\ \cdot \qquad\qquad\qquad\qquad & \quad \cdot \\ \cdot \qquad\qquad\qquad\qquad & \quad \cdot \\ \cdot \qquad\qquad\qquad\qquad & \quad \cdot \\ k_{n1}x_1 + k_{n2}x_2 \quad \cdots \quad + k_{nn}x_n &= \lambda x_n \end{aligned}$$

which may be rearranged to

$$\begin{aligned} (k_{11} - \lambda)x_1 + k_{12}x_2 \quad \cdots \quad + k_{1n}x_n &= 0 \\ k_{21}x_1 + (k_{22} - \lambda)x_2 \quad \cdots \quad + k_{2n}x_n &= 0 \\ \cdot \qquad\qquad\qquad\qquad & \quad \cdot \\ \cdot \qquad\qquad\qquad\qquad & \quad \cdot \\ \cdot \qquad\qquad\qquad\qquad & \quad \cdot \\ k_{n1}x_1 + \qquad\qquad \cdots \quad + (k_{nn} - \lambda)x_n &= 0 \end{aligned}$$

This is a set of n homogeneous, linear equations in n unknowns, and such a set always has the solution $x_1 = x_2 = \cdots = x_n = 0$ no matter what the value of λ. The null vector is thus an eigenvector of every matrix with arbitrary eigenvalue. In practice, however, we are never interested in this solution, called the *trivial solution*. We always require solutions other than the trivial one, and by a well known theorem these will exist only if λ is chosen so as to set the determinant of coefficients of these equations equal to zero:

$$|k_{ij} - \lambda\delta_{ij}| = 0 \qquad \text{or} \qquad |K - \lambda I| = 0$$

This determinantal equation is called the *secular equation* of the eigenvalue problem and serves to determine the eigenvalues λ. When the secular determinant $|K - \lambda I|$ is expanded, it turns out to be a polynomial of degree n in λ called the *characteristic polynomial*. Because a polynomial of degree n can have at most n distinct roots, there are at most n distinct eigenvalues $\lambda_1, \lambda_2, \ldots, \lambda_n$ which are roots of the characteristic polynomial.

Some examples: (1) Find the eigenvalues of

$$A = \begin{pmatrix} 2 & 0 \\ 0 & \frac{1}{2} \end{pmatrix}$$

We construct the determinant $|A - \lambda I|$ and set it equal to zero:

$$\begin{vmatrix} 2 - \lambda & 0 \\ 0 & \frac{1}{2} - \lambda \end{vmatrix} = (2 - \lambda)(\tfrac{1}{2} - \lambda) = 0$$

The polynomial $(2 - \lambda)(\frac{1}{2} - \lambda) = \lambda^2 - (\frac{5}{2})\lambda + 1$ is evidently the characteristic polynomial with roots $\lambda_1 = 2$ and $\lambda_2 = \frac{1}{2}$ as eigenvalues.

(2) Find the eigenvalues of

$$K = \begin{pmatrix} 6 & 2 \\ 2 & 3 \end{pmatrix}$$

Construct

$$|K - \lambda I| = \begin{vmatrix} 6 - \lambda & 2 \\ 2 & 3 - \lambda \end{vmatrix} = \lambda^2 - 9\lambda + 14 = (\lambda - 2)(\lambda - 7) = 0$$

Hence $\lambda_1 = 7$; $\lambda_2 = 2$.

In the examples above the eigenvalues found were all real numbers and they were distinct, with no repeated roots. Matrices can have complex numbers as eigenvalues, but such matrices will not occur in our work. Matrices with repeated eigenvalues, that is, with characteristic polynomials that have multiple roots do, however, occur. Such repeated eigenvalues are said to be *degenerate*.

PROBLEM

1.12 Find the eigenvalues of

$$\begin{pmatrix} 3 & 1 & 1 \\ 1 & 0 & 2 \\ 1 & 2 & 0 \end{pmatrix}$$

1.19 EIGENVECTORS

To each root λ_j of the characteristic polynomial there corresponds an eigenvector. To find the components of the eigenvector corresponding to, say, λ_1, put $\lambda = \lambda_1$ in our set of homogeneous, linear equations $K\mathbf{x} = \lambda_1\mathbf{x}$ and solve for the components of $\mathbf{x}$. For the matrix

$$A = \begin{pmatrix} 2 & 0 \\ 0 & \frac{1}{2} \end{pmatrix}$$

the calculation is

$$(2 - \lambda_1)x_1 + (0)x_2 = 0$$
$$(0)x_1 + (\tfrac{1}{2} - \lambda_1)x_2 = 0$$

or, since $\lambda_1 = 2$,

$$(0)x_1 + (0)x_2 = 0$$
$$(0)x_1 + (-\tfrac{3}{2})x_2 = 0$$

It follows from these equations that they are satisfied by $x_2 = 0$ and $x_1 =$ anything. The result is typical, and it can be shown from the general theory of linear equations that the eigenvector corresponding to a nondegenerate eigenvalue is determined only to within an arbitrary component; hence if one component is fixed arbitrarily, the others are uniquely determined. Geometrically, the eigenvector corresponding to a nondegenerate eigenvalue is fixed as to its direction but not as to its length. Because the length is arbitrary, it is conventional (but not essential) to normalize the eigenvector to unit length. Thus in our example normalization requires $x_1^2 + x_2^2 = 1$; but $x_2 = 0$, so that $x_1^2 = 1$, and I take the positive square root $x_1 = 1$, although $x_1 = -1$ would be just as good. The complete, normalized eigenvector corresponding to $\lambda_1 = 2$ is therefore

$$\mathbf{x}_1 = \begin{pmatrix} 1 \\ 0 \end{pmatrix}$$

In a similar way one finds for the eigenvalue $\lambda_2 = \frac{1}{2}$ a normalized eigenvector

$$\mathbf{x}_2 = \begin{pmatrix} 0 \\ 1 \end{pmatrix}$$

The student should turn back to Figure 1.5 and locate these eigenvectors in the linear transformation performed by A on the vector space of two dimensions.

Another example:

$$K = \begin{pmatrix} 6 & 2 \\ 2 & 3 \end{pmatrix}; \qquad \lambda_1 = 7; \qquad \lambda_2 = 2$$

For $\lambda = \lambda_1$,

$$(6 - 7)x_1 + 2x_2 = 0$$
$$2x_1 + (3 - 7)x_2 = 0$$

which reduces to a single equation:

$$-x_1 + 2x_2 = 0$$

If we want the eigenvector to be normalized, we must also have $x_1^2 + x_2^2 = 1$, so that the resulting eigenvector is

$$\mathbf{x}_1 = \begin{pmatrix} 2/\sqrt{5} \\ 1/\sqrt{5} \end{pmatrix}$$

Similarly for $\lambda_2 = 2$, we have a normalized eigenvector

$$\mathbf{x}_2 = \begin{pmatrix} -1/\sqrt{5} \\ 2/\sqrt{5} \end{pmatrix}$$

It is extremely suggestive to follow these operations geometrically. Consider Figure 1.10. Before transformation of the space by K, the eigenvectors $\mathbf{x}_1$ and $\mathbf{x}_2$ are of unit length. After transformation with K, $\mathbf{x}_1$ becomes $7\mathbf{x}_1$ and $\mathbf{x}_2$ becomes $2\mathbf{x}_2$. If the student experiments with other unit vectors, he will find that the unit circle is transformed by K into an ellipse with principal axes oriented along the directions $\mathbf{x}_1$ and $\mathbf{x}_2$.

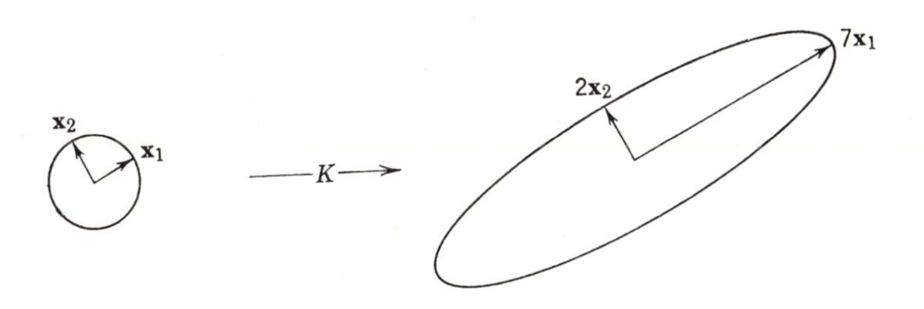

Figure 1.10 A linear transformation showing eigenvectors.

PROBLEM

1.13 Find the eigenvalues and normalized eigenvectors of the matrix

$$\begin{pmatrix} 1 & -6 \\ -6 & -4 \end{pmatrix}$$

1.20 DEGENERATE EIGENVALUES

We have just observed that the eigenvectors of a nondegenerate eigenvalue are determined to within an arbitrary component. Geometrically, this means that we know the line through the origin on which the vector lies, and any vector on this line is an eigenvector. It can be shown from the theory of linear equations that if an eigenvalue is doubly degenerate, two of the components of its eigenvector may be chosen arbitrarily; geometrically this means that the eigenvector is known to lie in a certain subspace of two dimensions, and any vector in this subspace is an eigenvector, all of them belonging to the same eigenvalue. Similarly the eigenvectors belonging to a triply degenerate eigenvalue all lie within a three dimensional subspace, and so on.

To illustrate these properties, the dilation matrix

$$D = \begin{pmatrix} 2 & 0 \\ 0 & 2 \end{pmatrix}$$

is doubly degenerate, for the secular equation

$$\begin{vmatrix} 2-\lambda & 0 \\ 0 & 2-\lambda \end{vmatrix} = (2-\lambda)^2 = 0$$

has roots $\lambda = 2, 2$. When we try to find the components of an eigenvector for $\lambda = 2$, we have to solve the linear equations

$$(2-\lambda)x_1 = 0$$

$$(2-\lambda)x_2 = 0$$

which for $\lambda = 2$ tell us only that x_1 and x_2 are arbitrary. But this makes very good geometric sense, for Figure 1.4 shows that *all* vectors in the x_1, x_2 plane are sent into multiples of themselves when operated on by D. This is just another way of saying that all vectors in the x_1, x_2 plane are eigenvectors of D.

1.21 TRANSPOSED MATRICES

The matrix K^T which is formed from K by rearranging the columns of K into the rows of K^T is known as the *transpose* of K. Symbolically $(k_{ij})^T = (k_{ji})$. For example, the transpose of

$$\begin{pmatrix} 1 & 2 & 3 \\ 0 & -1 & 4 \\ 0 & 0 & 6 \end{pmatrix}$$

is

$$\begin{pmatrix} 1 & 0 & 0 \\ 2 & -1 & 0 \\ 3 & 4 & 6 \end{pmatrix}$$

Up to now we have not specified in our notation for vectors whether $\mathbf{x}$ is to be written as a row or as a column. We henceforth specify $\mathbf{x}$ to be a column and $\mathbf{x}^T$ to be the same vector written as a row. Thus $\mathbf{x}^T$ is the transpose of $\mathbf{x}$. In terms of this notation and the ordinary rules of matrix multiplication, the scalar product of two vectors $\mathbf{x}$ and $\mathbf{y}$ may be written $\mathbf{x} \cdot \mathbf{y} = \mathbf{x}^T\mathbf{y}$.

An important theorem concerning transposed matrices states that the transpose of a product is the product of the transposes taken in reverse order. This theorem is true for the scalar product of two vectors, for the product of a matrix times a vector, and for the product of a matrix times a matrix. Symbolically,

$$\mathbf{x}^T\mathbf{y} = \mathbf{y}^T\mathbf{x}; \qquad (G\mathbf{x})^T = \mathbf{x}^T G^T; \qquad (AB)^T = B^T A^T \tag{1.2}$$

As an example of the second of these, if

$$G = \begin{pmatrix} 1 & 0 \\ 1 & 3 \end{pmatrix} \quad \text{and} \quad \mathbf{x} = \begin{pmatrix} 1 \\ 0 \end{pmatrix}$$

then

$$G\mathbf{x} = \begin{pmatrix} 1 & 0 \\ 1 & 3 \end{pmatrix}\begin{pmatrix} 1 \\ 0 \end{pmatrix} = \begin{pmatrix} 1 \\ 1 \end{pmatrix}$$

$$\mathbf{x}^T G^T = (1, 0)\begin{pmatrix} 1 & 1 \\ 0 & 3 \end{pmatrix} = (1, 1) = (G\mathbf{x})^T$$

A matrix K for which $K = K^T$ is said to be *symmetric*. Such matrices have their components symmetrically disposed with respect to the main diagonal, and they will play a role of peculiar importance in our work. The matrix A of Section 1.8 and the matrix K of Section 1.19 are examples of symmetric matrices.

The orthogonal matrix introduced in Section 1.16 has the property that its transpose is equal to its inverse $M^T = M^{-1}$ or $M^T M = I$, as the student may check by experimentation with the matrix R of Section 1.8.

1.22 EIGENVECTORS OF A SYMMETRIC MATRIX

A very important theorem states that the eigenvectors of a real, symmetric matrix which are associated with distinct eigenvalues are orthogonal. *Proof*: Let K be a real, symmetric matrix with two distinct eigenvalues λ_1 and λ_2 and let the eigenvectors associated with these eigenvalues be $\boldsymbol{\psi}_1$ and $\boldsymbol{\psi}_2$. Then

$$K\boldsymbol{\psi}_1 = \lambda_1 \boldsymbol{\psi}_1$$

$$K\boldsymbol{\psi}_2 = \lambda_2 \boldsymbol{\psi}_2$$

with $\lambda_1 \neq \lambda_2$. Form the scalar product of the first equation with $\boldsymbol{\psi}_2$ and the scalar product of the second equation with $\boldsymbol{\psi}_1$.

$$\begin{aligned} \boldsymbol{\psi}_2^T K \boldsymbol{\psi}_1 &= \lambda_1 \boldsymbol{\psi}_2^T \boldsymbol{\psi}_1 \\ \boldsymbol{\psi}_1^T K \boldsymbol{\psi}_2 &= \lambda_2 \boldsymbol{\psi}_1^T \boldsymbol{\psi}_2 \end{aligned} \tag{1.3}$$

Now transpose both sides of the second part of (1.3), using the rule (1.2) for the transpose of a product, and subtract the result from the first of equations 1.3:

$$\boldsymbol{\psi}_2^T K \boldsymbol{\psi}_1 - \boldsymbol{\psi}_2^T K^T \boldsymbol{\psi}_1 = \lambda_1 \boldsymbol{\psi}_2^T \boldsymbol{\psi}_1 - \lambda_2 \boldsymbol{\psi}_2^T \boldsymbol{\psi}_1$$

or

$$\boldsymbol{\psi}_2^T (K - K^T) \boldsymbol{\psi}_1 = (\lambda_1 - \lambda_2) \boldsymbol{\psi}_2^T \boldsymbol{\psi}_1$$

Now if $K = K^T$, then the left hand side is zero, and there remains

$$(\lambda_1 - \lambda_2)\boldsymbol{\psi}_2{}^T\boldsymbol{\psi}_1 = 0$$

But we are given that $\lambda_1 \neq \lambda_2$, and it follows that we must have $\boldsymbol{\psi}_2{}^T\boldsymbol{\psi}_1 = 0$, meaning that $\boldsymbol{\psi}_1$ and $\boldsymbol{\psi}_2$ are orthogonal. Note that this is not necessarily the case if both $\boldsymbol{\psi}_1$ and $\boldsymbol{\psi}_2$ belong to a degenerate eigenvalue $\lambda_1 = \lambda_2$.

The student should confirm for himself the algebraic and geometric validity of this theorem using as examples the eigenvectors of A and K in Section 1.19 together with an examination of Figures 1.5 and 1.10.

PROBLEMS

1.14 Find the eigenvalues and normalized eigenvectors of

$$\begin{pmatrix} 7 & -2 & 0 \\ -2 & 6 & -2 \\ 0 & -2 & 5 \end{pmatrix}$$

1.15 For K a symmetric matrix with eigenvalues λ_j and n orthonormal eigenvectors $\mathbf{E}_j$, calculate the matrix C of order n whose elements are $c_{ij} = \mathbf{E}_i{}^T K \mathbf{E}_j$.

1.16 Prove that if K is nonsingular and $K\mathbf{x} = \lambda\mathbf{x}$, then $\mathbf{x}$ is also an eigenvector of K^{-1} with eigenvalue λ^{-1}.

1.23 EIGENVALUES OF A SYMMETRIC MATRIX

Another very important theorem states that the eigenvalues of a real, symmetric matrix are all real. To prove this, let the operation of taking the complex conjugate of a scalar or a vector be denoted by an asterisk (*), so that λ^* is the complex conjugate of λ and $\boldsymbol{\psi}^*$ is the complex conjugate of $\boldsymbol{\psi}$. Then

$$K\boldsymbol{\psi} = \lambda\boldsymbol{\psi}$$

and if we take the complex conjugate of each side of this equation

$$K\boldsymbol{\psi}^* = \lambda^*\boldsymbol{\psi}^*$$

where K, being defined real, is unaffected by taking its complex conjugate. Now from our previous theorem, if $\lambda \neq \lambda^*$, then $\boldsymbol{\psi}^*$ must be orthogonal to $\boldsymbol{\psi}$. With the exception of the trivial solution $\boldsymbol{\psi} = \mathbf{0}$, however, it is impossible for a vector to be orthogonal to its own complex conjugate (try it!), and it must follow that $\lambda = \lambda^*$, which is to say that λ is real. The student may test the truth of these theorems on the examples of Section 1.19.

1.24 RECAPITULATION

We can summarize the importance of these theorems by noting that if all of the eigenvalues of a real, symmetric matrix are distinct, there will be n in number and to each one is assigned an eigenvector that is orthogonal to every other eigenvector. If we by convention normalize these eigenvectors, then they form an orthonormal set in the space of n dimensions and are thus a suitable basis (or coordinate system) in the space of n dimensions. Hence whenever a symmetric matrix K occurs in a physical problem, the computational details of that problem usually become enormously simplified if the coordinate axes are chosen to lie along the eigenvectors of K. If the eigenvalues are not all distinct, they may be grouped into degenerate sets, and every eigenvector in a given degenerate set will be automatically orthogonal to every eigenvector belonging to another degenerate set. Within the members of a given degenerate set, however, the eigenvectors are not necessarily orthogonal. We have seen that they all lie within some subspace of the vector space of n dimensions, the dimension of the subspace being equal to the degeneracy of the eigenvalue, so that the eigenvectors of a doubly degenerate eigenvalue lie within a plane, the eigenvectors of a triply degenerate eigenvalue lie within a subspace (volume) of three dimensions, and so on.

So long as an eigenvector belonging to a degenerate eigenvalue lies within its assigned subspace, its orientation within that subspace is arbitrary. It is usually convenient, however, to choose these degenerate eigenvectors to be orthonormal, so that they form a coordinate system within the assigned subspace. No set of orthonormal, degenerate eigenvectors is, however, to be preferred over any other *ab initio;* which set is chosen usually depends on other features of the problem in hand.

In any case, once subsets of orthonormal axes have been chosen for every degenerate eigenvalue, the entire vector space of n dimensions is spanned by n orthonormal eigenvectors, and what I said previously still stands: the eigenvectors of a real, symmetric matrix may be taken to be a basis in the vector space of n dimensions.

PROBLEM

1.17 Prove that if λ is degenerate, then any linear combination of its degenerate eigenvectors is itself an eigenvector with eigenvalue λ.

1.25 A DEGENERATE EIGENVALUE PROBLEM

Let us analyze in detail a specific example. Find the eigenvalues and a set of normalized eigenvectors ψ_k for

$$K = \begin{pmatrix} 5 & \sqrt{2} & 1 \\ \sqrt{2} & 6 & \sqrt{2} \\ 1 & \sqrt{2} & 5 \end{pmatrix} = K^T$$

The characteristic polynomial $f(\lambda) = |K - \lambda I|$ turns out to be

$$\begin{aligned} -f(\lambda) &= \lambda^3 - 16\lambda^2 + 80\lambda - 128 \\ &= (\lambda - 4)^2(\lambda - 8) \end{aligned}$$

Hence the eigenvalues are $\lambda_1 = \lambda_2 = 4$; $\lambda_3 = 8$. The third normalized eigenvector corresponding to the nondegenerate λ_3 is easily obtained. Its components satisfy

$$\begin{aligned} (5 - 8)x_1 + (\sqrt{2})x_2 + (1)x_3 &= 0 \\ (\sqrt{2})x_1 + (6 - 8)x_2 + (\sqrt{2})x_3 &= 0 \\ x_1^2 + x_2^2 + x_3^2 &= 1 \end{aligned}$$

whence $\psi_3 = (\frac{1}{2}, \sqrt{2}/2, \frac{1}{2})$. For $\lambda_1 = \lambda_2 = 4$, however, the set of three homogeneous, linear equations reduces to a single equation

$$x_1 + \sqrt{2}\, x_2 + x_3 = 0 \tag{1.4}$$

and both eigenvectors satisfy this equation together with the normalization condition

$$x_1^2 + x_2^2 + x_3^2 = 1$$

If we pick any arbitrary value for one of the components, then the other two are determined; thus, for example, if $x_2 = 0$, then (1.4) becomes $x_1 + x_3 = 0$, and a possible eigenvector is $\psi_1 = (\sqrt{2}/2, 0, -\sqrt{2}/2)$, which, we note in passing, is orthogonal to ψ_3. Another eigenvector ψ_2 associated with the eigenvalue 4 is obtained by making its components satisfy equation 1.4, and for convenience we also choose ψ_2 to be orthogonal to ψ_1:

$$\psi_2^T\psi_1 = (\sqrt{2}/2)x_1 - (\sqrt{2}/2)x_3 = 0$$

When normalized, the result is $\psi_2 = (\frac{1}{2}, -\sqrt{2}/2, \frac{1}{2})$.

A clearer picture of the meaning of these operations is obtained if we construct the corresponding linear transformation in the vector space of three dimensions. If K operates on all unit vectors (those with their ends resting on the unit sphere), the eigenvector ψ_3 is lengthened by a factor 8, and *all* unit vectors in the subspace spanned by ψ_1, ψ_2 are lengthened by a factor 4 (Figure 1.11). The unit sphere is thus elongated into an ellipsoid retaining its symmetry in the 1, 2 plane. Had our matrix been nondegenerate, the sphere would have been converted into an ellipsoid with all three of its principal axes distinct. The geometric meaning of the degenerate subspace spanned by the degenerate eigenvectors ψ_1 and ψ_2 is thus clear, for there is no unique way of drawing orthogonal axes within it; and the choice of a coordinate system is to this extent indeterminate.

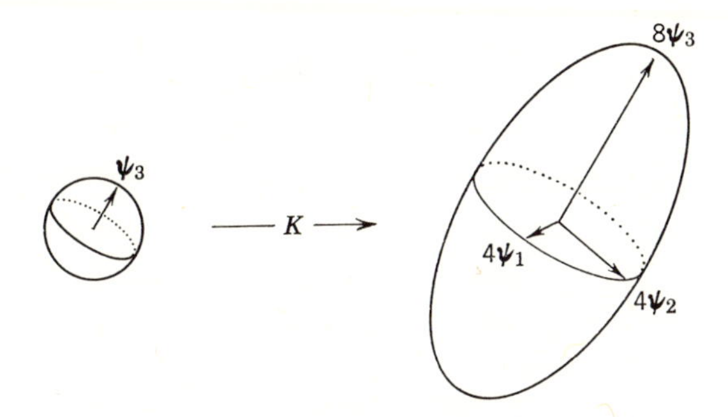

Figure 1.11 A degenerate linear transformation showing eigenvectors.

The student should correlate these pictures with Figure 1.4 for the degenerate dilation matrix D, which possesses no unique principal axes after the fashion of matrix A of Figure 1.5. It is evident that the subspace spanned by the degenerate eigenvectors of a symmetric matrix is locally dilated by the linear transformation.

PROBLEMS

1.18 Construct other pairs of orthonormal, degenerate eigenvectors of K as linear combinations of the pair ψ_1, ψ_2 of Section 1.25.

1.19 Find the eigenvalues and a set of orthonormal eigenvectors of the degenerate matrix

$$\begin{pmatrix} -7 & 4 & -4 \\ 4 & -1 & -8 \\ -4 & -8 & -1 \end{pmatrix}$$

1.26 EIGENVECTORS OF COMMUTING, SYMMETRIC MATRICES

If A and B commute, $AB = BA$, they possess a complete set of eigenvectors in common. Thus suppose that

$$A\psi = \lambda\psi$$

Operate on both sides by B:

$$BA\psi = \lambda B\psi$$

But $BA = AB$, so that

$$AB\psi = \lambda B\psi$$

or

$$A(B\psi) = \lambda(B\psi)$$

This states that $B\psi$ is an eigenvector of A. But ψ is also an eigenvector of A, and if λ is nondegenerate, then $B\psi$ can only be some multiple of ψ,

$$B\psi = \mu\psi$$

which is to say that ψ is also an eigenvector of B. This will be true for every nondegenerate eigenvector of A, and all such eigenvectors are therefore shared in common by A and B.

If λ is doubly degenerate, then there exist two normalized, mutually orthogonal vectors ψ_1 and ψ_2 for which

$$A\psi_1 = \lambda\psi_1; \qquad A\psi_2 = \lambda\psi_2 \tag{1.5}$$

We note furthermore from Section 1.20 and Problem 1.17 that any linear combination of ψ_1 and ψ_2 is also an eigenvector of A:

$$\begin{aligned} A(c_1\psi_1 + c_2\psi_2) &= c_1A\psi_1 + c_2A\psi_2 = c_1\lambda\psi_1 + c_2\lambda\psi_2 \\ &= \lambda(c_1\psi_1 + c_2\psi_2) \end{aligned}$$

Now let B operate on both sides of the first of equations 1.5, remembering that B commutes with A,

$$BA\psi_1 = \lambda B\psi_1$$
$$A(B\psi_1) = \lambda(B\psi_1)$$

so that $B\psi_1$ is also an eigenvector of A. But from the discussion above the most that can be said is that this implies

$$B\psi_1 = c_1\psi_1 + c_2\psi_2$$

with c_1 and c_2 unknown. Only for the special case $c_2 = 0$ could ψ_1 be taken as an eigenvector of B.

On the other hand, the original choice of the basis pair ψ_1, ψ_2 was to an extent arbitrary; the question now is, could we by another choice of basis pair find a set of common eigenvectors for A and B? That is, could we choose c_1 and c_2 in such a way as to make not ψ_1 and ψ_2 individually, but the linear combination $c_1\psi_1 + c_2\psi_2$ an eigenvector of B?

$$B(c_1\psi_1 + c_2\psi_2) = \mu(c_1\psi_1 + c_2\psi_2)? \tag{1.6}$$

To answer this question we take the scalar products of both sides of equation 1.6 with ψ_1 and with ψ_2,

$$\begin{aligned} c_1(\psi_1{}^T B\psi_1) + c_2(\psi_1{}^T B\psi_2) &= \mu c_1 \\ c_1(\psi_2{}^T B\psi_1) + c_2(\psi_2{}^T B\psi_2) &= \mu c_2 \end{aligned} \tag{1.7}$$

in which the right hand side has been simplified by virtue of the orthonormality of ψ_1 and ψ_2. The four scalar quantities

$$H_{ij} = \psi_i{}^T B\psi_j \tag{1.8}$$

are the components of the 2×2 matrix of coefficients of a pair (1.7) of linear equations in c_1 and c_2. From the rule on transposition of a product (equation 1.2)

$$(\psi_1{}^T B\psi_2)^T = \psi_2{}^T B^T \psi_1$$

But $B = B^T$ is symmetric, so that the matrix of coefficients (1.8) is also symmetric with $H_{12} = H_{21}$. Rearranging equations 1.7,

$$\begin{aligned} (H_{11} - \mu)c_1 + H_{12}c_2 &= 0 \\ H_{21}c_1 + (H_{22} - \mu)c_2 &= 0 \end{aligned}$$

which is a set of two homogeneous, linear equations in two unknown c's. In the usual way, such a set can have a nontrivial solution only if the determinant of coefficients vanishes

$$\begin{vmatrix} H_{11} - \mu & H_{12} \\ H_{21} & H_{22} - \mu \end{vmatrix} = 0 \tag{1.9}$$

The roots μ of this equation give two of the eigenvalues of B, and its two normalized eigenvectors each give a pair of coefficients c_1, c_2 which will make the linear combination $c_1\psi_1 + c_2\psi_2$ a simultaneous, normalized eigenvector of both A and B.

If λ is n fold degenerate, the secular equation to be solved is of the nth order.

$$|H_{ij} - \mu\delta_{ij}| = 0$$

whose n roots μ belong to B and whose n orthonormal eigenvectors $(c_{k1}, c_{k2}, \ldots, c_{kn})$ constitute the sets of coefficients necessary to construct n linear combinations,

$$c_{k1}\psi_1 + c_{k2}\psi_2 + \cdots + c_{kn}\psi_n$$

which are simultaneously eigenvectors of both A and B.

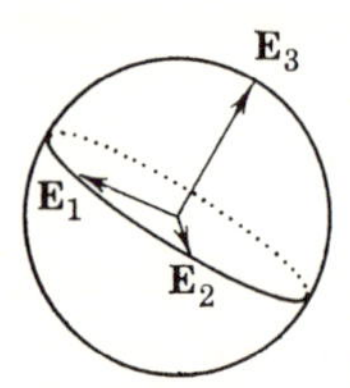

Figure 1.12 Eigenvectors before linear transformation.

As in Section 1.25, a few geometric diagrams may help clarify the meaning of this theorem. Suppose that we have a nondegenerate matrix G with eigenvalues λ_k and eigenvectors $\mathbf{E}_k$ which commutes with the degenerate K of Section 1.25. Our theorem then states that G and K have eigenvectors in common. Because G is nondegenerate, the orientation of its eigenvectors is

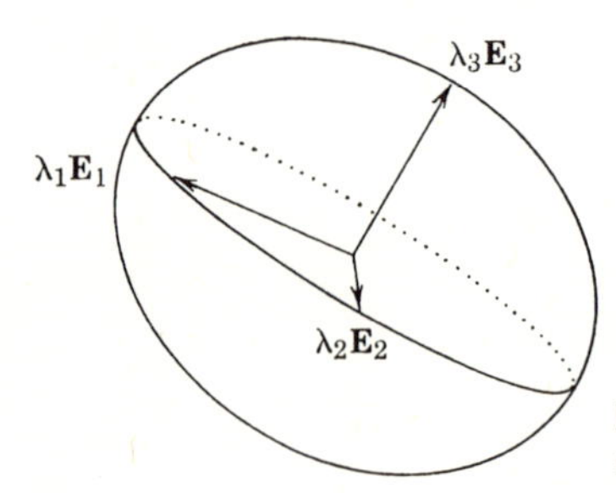

Figure 1.13 Eigenvectors after linear transformation.

uniquely determined, say as in Figure 1.12, so that according to our theorem, $\mathbf{E}_3$ automatically coincides with the nondegenerate ψ_3. Let us picture the result of the linear transformation G on the unit sphere to be the flattened ellipsoid of Figure 1.13. By comparison of Figure 1.13 with Figure 1.11 for the linear transformation K, we perceive that while ψ_1 and ψ_2 do not necessarily coincide with $\mathbf{E}_1$ and $\mathbf{E}_2$, they can be made to do so by a rotation of coordinate axes (a change of basis) in the K diagram. This is permissible, for any vector lying in the subspace 1, 2 of the K diagram is an eigenvector of K. It is the rotation of these axes (i.e., the construction of an orthogonal transformation), which is the result of the solution of the secular equation 1.9.

1.27 VECTOR SPACES WITH COMPLEX NUMBERS

So far we have considered only vectors and matrices whose components are real numbers. All of our theorems generalize to the case where complex numbers are admitted. The scalar product of two vectors is written $\mathbf{x}^\dagger\mathbf{y}$, where $\mathbf{x}^\dagger$ is written as a row and $\mathbf{y}$ as a column, but in addition to transposing $\mathbf{x}$ to form $\mathbf{x}^\dagger$ we also take its complex conjugate. With this definition $\mathbf{x}^\dagger\mathbf{x}$ is always positive real, so that the norms of complex vectors are always positive real.

In a similar way the transposition of a matrix is always accompanied by taking the complex conjugate of every component. Thus for complex matrices, $K^\dagger$ means transposition of K accompanied by complex conjugation. If $K = K^\dagger$, the matrix is said to be *Hermitian*, which is a generalization of the idea of symmetry for real matrices. As for symmetric, real matrices, the following theorems exist for Hermitian matrices:

1. The eigenvalues of an Hermitian matrix are real.
2. The eigenvectors of an Hermitian matrix belonging to different eigenvalues are orthogonal.
3. Hermitian matrices that commute possess a complete set of eigenvectors in common.

The role of an orthogonal matrix in real matrix theory is played in complex matrix theory by the *unitary* matrix, which has the property $M^\dagger = M^{-1}$, or $M^\dagger M = I$. The student should verify by actual calculation that for $i = \sqrt{-1}$ and

$$M = \begin{pmatrix} 1/\sqrt{2} & 1/\sqrt{2} \\ -i/\sqrt{2} & i/\sqrt{2} \end{pmatrix} \tag{1.10}$$

then

$$M^\dagger = \begin{pmatrix} 1/\sqrt{2} & i/\sqrt{2} \\ 1/\sqrt{2} & -i/\sqrt{2} \end{pmatrix}$$

and

$$M^\dagger M = \begin{pmatrix} 1/\sqrt{2} & i/\sqrt{2} \\ 1/\sqrt{2} & -i/\sqrt{2} \end{pmatrix} \begin{pmatrix} 1/\sqrt{2} & 1/\sqrt{2} \\ -i/\sqrt{2} & i/\sqrt{2} \end{pmatrix} = \begin{pmatrix} 1 & 0 \\ 0 & 1 \end{pmatrix}$$

PROBLEM

1.20 Given

$$A = \begin{pmatrix} 7 & -2 & 0 \\ -2 & 6 & -2 \\ 0 & -2 & 5 \end{pmatrix} \quad \text{and} \quad B = \begin{pmatrix} 4 & 2 & 2 \\ 2 & 7 & 4 \\ 2 & 4 & 7 \end{pmatrix}$$

(1) Show that A and B commute.

(2) Find a set of eigenvectors common to both A and B.

REFERENCES

1. F. B. Hildebrand, *Methods of Applied Mathematics*, Second Edition, Prentice-Hall, Inc., Englewood Cliffs, N.J., 1965, Chapter 1.
2. C. Lanczos, *Applied Analysis*, Prentice-Hall, Inc., Englewood Cliffs, N.J., 1956.
3. R. Courant and D. Hilbert, *Methods of Mathematical Physics*, Volume 1, Interscience, New York, 1953, Chapter 1.
4. H. Margenau and G. M. Murphy, *The Mathematics of Physics and Chemistry*, Second Edition, D. Van Nostrand, Inc., Princeton, 1956, Chapter 4.

Chapter 2

HÜCKEL MOLECULAR ORBITAL THEORY

2.1 HÜCKEL DIAGRAMS

We already have enough mathematical apparatus to construct a poor man's quantum mechanics for planar, aromatic hydrocarbons. While the methods employed bear only a superficial resemblance to the more sophisticated quantum mechanics that we develop later, the results are nevertheless capable of reproducing approximately many of the properties of these compounds which are familiar to the organic chemist.

We start with a model (Hückel diagram) of the hydrocarbon consisting of carbon atoms (circles) and bonds (lines) drawn in such a way as to reproduce the molecular structure formulas of the organic chemist. Some samples are shown in Figure 2.1, and we shall later learn that the bonds are to be identified with the π bonds of a conjugated, organic molecule.

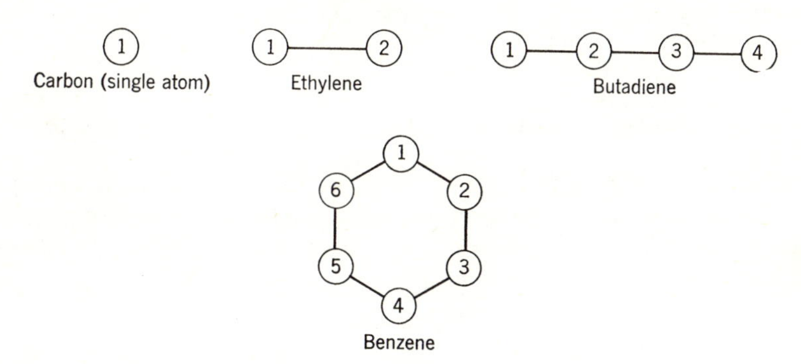

Figure 2.1 Some Hückel diagrams.

2.2 SIMPLIFIED QUANTUM POSTULATES

The atoms in the Hückel diagram are numbered in any arbitrary way. Into the atomic framework we feed electrons, and we postulate that the electrons are distributed over the atoms according to a probability law which may be calculated as follows:

1. The "state" of an electron shall be said to be determined by a *molecular orbital* defined to be a vector $\psi = (c_1, c_2, \ldots, c_n)$ in an n dimensional vector space where n is the number of carbon atoms in the compound. The components c_j of the vector have the interpretation that c_j^2 is the probability of observing the electron on the jth atom. Because the probability of observing the electron somewhere in the molecule is unity, we must have $c_1^2 + c_2^2 + \cdots + c_n^2 = 1$, or, in vector notation,

$$\psi^T\psi = 1$$

2. The acceptable vectors that may be used to describe the state of an electron are the eigenvectors of an n dimensional matrix H called the Hamiltonian matrix. This matrix is to be constructed in such a way that $H_{ii} = \alpha$; $H_{ij} = \beta$ if $i \neq j$ and atoms i and j are bonded; $H_{ij} = 0$ otherwise. The parameters α and β have the dimensions of energy, so that eigenvalues E associated with eigenvectors ψ also have the dimensions of energy, and an eigenvalue is interpreted to be the energy of an electron in the state ψ. Note that H is symmetric $H_{ij} = H_{ji}$, so that the ψ_j will form an orthonormal set of vectors.

3. The exclusion principle: if a molecule possesses more than one electron distributed over its atoms, the same state vector may be used to describe the state of not more than two electrons at a time. Other electrons must then be described by other state vectors. Finally, the total electron density at atom j is obtained by summing at that atom the contributions c_j^2 from all electrons in the molecule.

2.3 THE CARBON ATOM

To get an idea of how these rules work, let us examine the simplest possible case, that of a single carbon atom (Figure 2.1). Because $n = 1$, our vector space is one dimensional, $\psi = c_1$, and H consists of the single element $H_{11} = \alpha$. The eigenvectors of H satisfy

$$H\psi = E\psi$$

or

$$\alpha c_1 = Ec_1$$

whence $E = \alpha$ is the single energy available to an electron in a carbon atom. The corresponding eigenvector after normalization is

$$\psi = 1$$

so that $c_1 = 1$, and the probability of observing the electron on the atom is $1^2 = 1$. All of this is trivially simple, but it does yield a useful interpretation of α, telling us that α is the energy of an electron in an isolated carbon atom. It is customary to take the zero of energy for quantum mechanical systems as that when all the particles involved are at infinite separation. With this convention α must be negative, for the carbon atom is electronically stable, and its bound electron must lie at an energy level lower than zero.

2.4 ETHYLENE

Referring to Figure 2.1, our vector space is of dimension $n = 2$, and we set up matrix elements

$$H_{11} = H_{22} = \alpha; \qquad H_{12} = H_{21} = \beta$$

Then

$$H\psi = E\psi$$

or

$$\begin{pmatrix} \alpha & \beta \\ \beta & \alpha \end{pmatrix}\begin{pmatrix} c_1 \\ c_2 \end{pmatrix} = E\begin{pmatrix} c_1 \\ c_2 \end{pmatrix}$$

The secular equation is

$$\begin{vmatrix} \alpha - E & \beta \\ \beta & \alpha - E \end{vmatrix} = 0$$

whence we have eigenvalues $E_1 = \alpha + \beta$; $E_2 = \alpha - \beta$. To obtain results corresponding to experiment, it turns out that we must empirically assume that β like α is also negative, so that the most stable energy level for an electron in ethylene is E_1, and the neutral molecule may accommodate both of its electrons in this level without violating the exclusion principle. Thus the total ground state energy for the neutral molecule is $2E_1 = 2\alpha + 2\beta$.

To compare this molecular energy with the atomic energy of two unbonded carbon atoms, rub out the bond between 1 and 2 and calculate the energy for the Hückel diagram:

① ②

The modified Hamiltonian matrix becomes

$$H = \begin{pmatrix} \alpha & 0 \\ 0 & \alpha \end{pmatrix}$$

whence the unbonded atoms possess a doubly degenerate energy level $E_1 = E_2 = \alpha$, so that two electrons in this level have a total energy 2α. Comparing this with the molecular energy of ethylene, it is evident that the bond energy (read π bond energy) in ethylene is 2β, which gives us a physical interpretation of β as one half the energy of an olefinic π bond.

Figure 2.2 illustrates the change in energy levels upon molecule formation in ethylene. The molecular energy levels are classified as bonding or antibonding according to whether they lie below or above the original atomic levels before molecule formation.

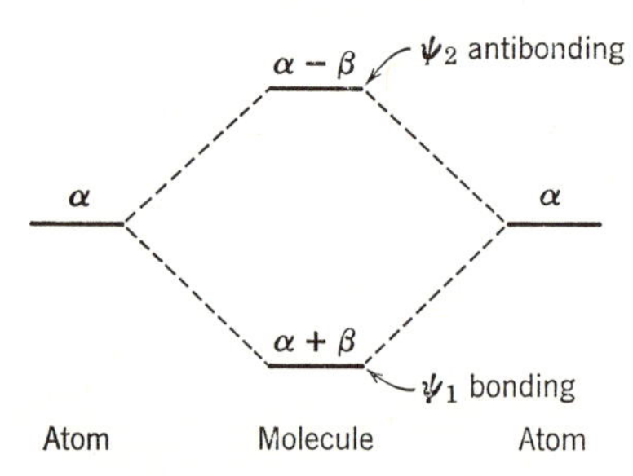

Figure 2.2 Comparison of atomic and molecular energy levels in ethylene.

Associated with E_1 in ethylene is an eigenvector $\psi_1 = (1/\sqrt{2}, 1/\sqrt{2})$ and with E_2 an eigenvector $\psi_2 = (1/\sqrt{2}, -1/\sqrt{2})$. Because both electrons in the ground state of the molecule are described by ψ_1, each electron makes a contribution of $(1/\sqrt{2})^2 = \frac{1}{2}$ to the total electron density at each carbon, whence there is a total of $\frac{1}{2} + \frac{1}{2} = 1$ electron distributed over each carbon atom. The electronic configuration of two electrons in ψ_1 is conventionally denoted by the symbol $(\psi_1)^2$.

Note that the same electron density at each carbon would be obtained from ψ_2, and at first sight this would appear to be inconsistent with the fact that ψ_1 possesses a lower energy. If, however, the electrons were conceived to be continuously distributed throughout the spacial region in the neighborhood of the atoms, then the change in sign between c_1 and c_2 in ψ_2 would imply a region between the atoms of zero electron density—a node as we shall later learn to call it. ψ_1 exhibits no such change in sign, and we may infer a nonvanishing electron density between the atoms. These conclusions are in accord with the chemist's traditional picture of the formation of a chemical bond as being accompanied by the buildup of electron density between the bonded atoms and constitute our empirical reason for taking β to be negative.

If one electron is in ψ_1 and the other is in ψ_2, we speak of the molecule as being in an excited state and write its configuration as $(\psi_1)(\psi_2)$, but according to our present formalism, this excited state would have a total energy $E_1 + E_2 = 2\alpha$, which is the same as two isolated carbon atoms. The excited state $(\psi_1)(\psi_2)$ would therefore be nonbonding. If both electrons are in state ψ_2, the total energy would be $2E_2 = 2\alpha - 2\beta$, which exceeds the energy of two isolated carbon atoms by the (positive) amount -2β. The state $(\psi_2)^2$ would therefore be antibonding and would in the absence of other bonding forces lead to dissolution of the molecule.

2.5 BUTADIENE

We have a four dimensional vector space and an H matrix with $H_{11} = H_{22} = H_{33} = H_{44} = \alpha$; $H_{12} = H_{23} = H_{34} = \beta$; $H_{13} = H_{14} = H_{24} = 0$. The secular equation is

$$\begin{vmatrix} \alpha - E & \beta & 0 & 0 \\ \beta & \alpha - E & \beta & 0 \\ 0 & \beta & \alpha - E & \beta \\ 0 & 0 & \beta & \alpha - E \end{vmatrix} = 0 \tag{2.1}$$

The roots of this equation are unaffected if β is factored out of each row, and it is also convenient to define $x = (\alpha - E)/\beta$. Then

$$\begin{vmatrix} x & 1 & 0 & 0 \\ 1 & x & 1 & 0 \\ 0 & 1 & x & 1 \\ 0 & 0 & 1 & x \end{vmatrix} = 0$$

Expansion of the determinant leads to a secular polynomial

$$x^4 - 3x^2 + 1 = 0$$

with roots

$$x = \pm(\tfrac{3}{2} \pm \tfrac{1}{2}\sqrt{5})^{1/2}$$

The allowed energy levels are therefore ($E = \alpha - x\beta$):

$$\begin{aligned} E_1 &= \alpha + 1.618\beta \\ E_2 &= \alpha + 0.618\beta \\ E_3 &= \alpha - 0.618\beta \\ E_4 &= \alpha - 1.618\beta \end{aligned} \tag{2.2}$$

The most stable configuration for the neutral molecule will arise if we place two electrons in each of the two lowest energy levels E_1 and E_2, thus leading to a total of $E = 2E_1 + 2E_2 = 4\alpha + 4.472\beta$. It is instructive to note that not only is this energy lower by a bond energy 4.472β than the energy of four isolated carbons,

but it is also lower by 0.472β than the energy of two isolated ethylenes with Hückel diagram

and secular equation

$$\begin{vmatrix} \alpha - E & \beta & 0 & 0 \\ \beta & \alpha - E & 0 & 0 \\ 0 & 0 & \alpha - E & \beta \\ 0 & 0 & \beta & \alpha - E \end{vmatrix} = \begin{vmatrix} \alpha - E & \beta \\ \beta & \alpha - E \end{vmatrix}^2 = 0$$

This increased stabilization energy given to a conjugated molecule by delocalization of its double bonds is called its *delocalization* energy.

Eigenvectors corresponding to each of the eigenvalues (2.2) are

$$\begin{aligned} \psi_1 &= (0.372, \quad 0.601, \quad 0.601, \quad 0.372) \\ \psi_2 &= (0.601, \quad 0.372, -0.372, -0.601) \\ \psi_3 &= (0.601, -0.372, -0.372, \quad 0.601) \\ \psi_4 &= (0.372, -0.601, \quad 0.601, -0.372) \end{aligned} \tag{2.3}$$

When the first two levels of butadiene are filled with two electrons each, we write the configuration $(\psi_1)^2(\psi_2)^2$, and there will be $2(0.372)^2 + 2(0.601)^2 = 1$ electron on the average on each carbon. There are no nodes or changes of sign in the components of ψ_1, which is characteristic of the most stable electronic states. The two electrons in ψ_2, however, each have a node between atoms 2 and 3, and from our experience with ethylene, this would suggest that some of the bonding between 2 and 3 which is provided by ψ_1 is canceled out by ψ_2, which at the same time reinforces the bonding between atoms 1 and 2 and atoms 3 and 4. In other words, butadiene should have more double bond character for the outer two bonds than for the inner one. This conclusion can be made quantitative by defining what is known as the *bond order* p_{rs} for the two bonded atoms r and s. This is computed by forming the products $c_r c_s$ taken from every occupied state vector of the molecule for the fixed positions r and s and summing the contributions from each electron.

Thus for ethylene, both electrons in the ground state are in $\psi_1 = (1/\sqrt{2}, 1/\sqrt{2})$. Hence $c_1 c_2 = (1/\sqrt{2})(1/\sqrt{2}) = \frac{1}{2}$ is the contribution from each electron, so that the total bond order is $\frac{1}{2} + \frac{1}{2} = 1$, and we may take this as the bond order of a true, olefinic double bond. For butadiene in the ground state, ψ_1 and ψ_2 are each occupied by two electrons. The bond between atoms 1 and 2 has a contribution $c_1 c_2 = (0.372)(0.601)$ from ψ_1 and a contribution $c_1 c_2 = (0.601)(0.372)$ from ψ_2. The total bond order for this bond is thus $2(0.372)(0.601) + 2(0.601)(0.372) = 0.894$, somewhat less than the ethylenic double bond.

Similarly, the bond order for the 2–3 bond is $2c_2c_3$ from ψ_1 plus $2c_2c_3$ from ψ_2, or $2(0.601)(0.601) + 2(0.372)(-0.372) = 0.447$—an even weaker double bond.

PROBLEMS

2.1 The allyl system has the structure

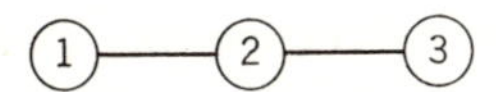

(1) Calculate the energy levels and Hückel molecular orbitals for allyl.
(2) Calculate the bond energy for the molecule and the charge density at each atom for the species:
 (a) The allyl cation (two electrons).
 (b) The allyl radical (three electrons).
 (c) The allyl anion (four electrons).

2.2 The hypothetical cyclobutadiene molecule has a Hückel diagram

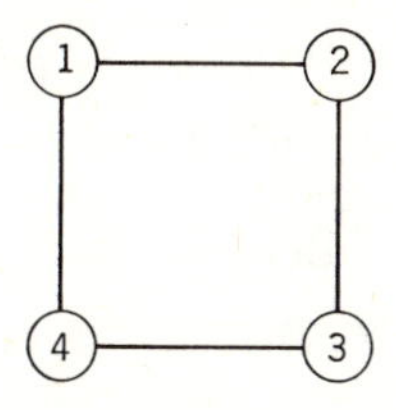

(1) Calculate the energy levels and Hückel molecular orbitals for cyclobutadiene.

(2) Calculate the delocalization energy for the molecule. On the basis of this calculation, can you explain why cyclobutadiene has never been synthesized?

2.3 Calculate the bond orders for the excited state $(\psi_1)^2(\psi_2)(\psi_3)$ of butadiene. What does this say about the olefinic character of photochemically excited butadiene?

2.6 BENZENE

Our experience by now is such that we may rapidly set up the secular determinant.

$$\begin{vmatrix} x & 1 & 0 & 0 & 0 & 1 \\ 1 & x & 1 & 0 & 0 & 0 \\ 0 & 1 & x & 1 & 0 & 0 \\ 0 & 0 & 1 & x & 1 & 0 \\ 0 & 0 & 0 & 1 & x & 1 \\ 1 & 0 & 0 & 0 & 1 & x \end{vmatrix} = 0$$

where, as before, $x = (\alpha - E)/\beta$. The roots prove to be $x = -2, -1, -1, 1, 1, 2$, and we note that for this molecule two of the energy levels are degenerate. In order of increasing energy, the eigenvalues are

$$E_1 = \alpha + 2\beta$$
$$E_2 = E_3 = \alpha + \beta$$
$$E_4 = E_5 = \alpha - \beta$$
$$E_6 = \alpha - 2\beta$$

and the eigenvectors prove to be

$$\begin{aligned} &\psi_1 = (1/\sqrt{6})(1, 1, 1, 1, 1, 1) \\ &\begin{cases} \psi_2 = (1/\sqrt{4})(0, 1, 1, 0, -1, -1) \\ \psi_3 = (1/\sqrt{12})(2, 1, -1, -2, -1, 1) \end{cases} \\ &\begin{cases} \psi_4 = (1/\sqrt{4})(1, -1, 0, 1, -1, 0) \\ \psi_5 = (1/\sqrt{12})(1, 1, -2, 1, 1, -2) \end{cases} \\ &\psi_6 = (1/\sqrt{6})(1, -1, 1, -1, 1, -1) \end{aligned} \tag{2.4}$$

in which I have bracketed the degenerate sets. Because of the degeneracy, this is only one of many possible sets of eigenvectors, and any normalized linear combination of ψ_2 and ψ_3 is also an eigenvector, as is any normalized linear combination of ψ_4 and ψ_5. Note that I have chosen the eigenvectors spanning the two dimensional degenerate subspaces to be mutually orthogonal, so that the entire set of six vectors is an orthonormal basis in the vector space of six dimensions.

The neutral benzene molecule will have six electrons to be accommodated in these energy levels, and our exclusion principle requires that in the ground state, two electrons will reside in each of ψ_1, ψ_2, and ψ_3. The total energy of the configuration $(\psi_1)^2(\psi_2)^2(\psi_3)^2$ will thus be $E = 2E_1 + 2E_2 + 2E_3 = 6\alpha + 8\beta$. Three isolated double bonds (cyclohexatriene) would have an energy $6\alpha + 6\beta$, and the difference 2β between the two is the delocalization energy of benzene.

To obtain the electron density at each atom, sum the squares of corresponding vector components over all electrons. Thus the electron density at atom 1 is $2[(1/\sqrt{6})^2 + 0^2 + (2/\sqrt{12})^2] = 1$, and the same value will be found for each atom in the molecule.

There is no difficulty in interpreting ψ_1 as being a bonding state vector, for it has no nodes anywhere in the molecule. ψ_2 and ψ_3 have two nodal points each, however, and upon inspection these prove to lie on two mutually perpendicular nodal lines passing through the molecule (Figure 2.3). Despite the antibonding character of these nodes, the energy $\alpha + \beta$ of an electron in either of ψ_2 or ψ_3 indicates that it adds overall stability to the molecule. It might at first appear that the location of these nodes would permit different atoms to have different properties. Should not, for example, atoms 1 and 4 which lie on a nodal line have properties different from atoms 2 and 3 or 5 and 6 which straddle a nodal line? The answer is no, for so long as they are mutually perpendicular, the nodal lines have an arbitrary orientation with respect to the molecule. Other orientations correspond to other pairs of mutually orthogonal vectors belonging to the eigenvalue $\alpha + \beta$. Thus ψ_2 and ψ_3 may be replaced by

$$\psi_2' = (1/\sqrt{4})(1, 1, 0, -1, -1, 0)$$
$$\psi_3' = (1/\sqrt{12})(1, -1, -2, -1, 1, 2)$$

All of our postulates are still satisfied, but the nodal lines now have the appearance of Figure 2.4. The quantum mechanical interpretation of this situation is that while we are permitted to know of the existence of a nodal line, the degeneracy of this energy level prohibits us from identifying its orientation.

I emphasize this arbitrariness by pointing out that ψ_2' and ψ_3' above are only one out of infinitely many orthonormal pairs of vectors that can be assigned to the degenerate level $\alpha + \beta$. In general, any pair of linear combinations

$$\psi_2' = a_{22}\psi_2 + a_{23}\psi_3$$
$$\psi_3' = a_{32}\psi_2 + a_{33}\psi_3$$

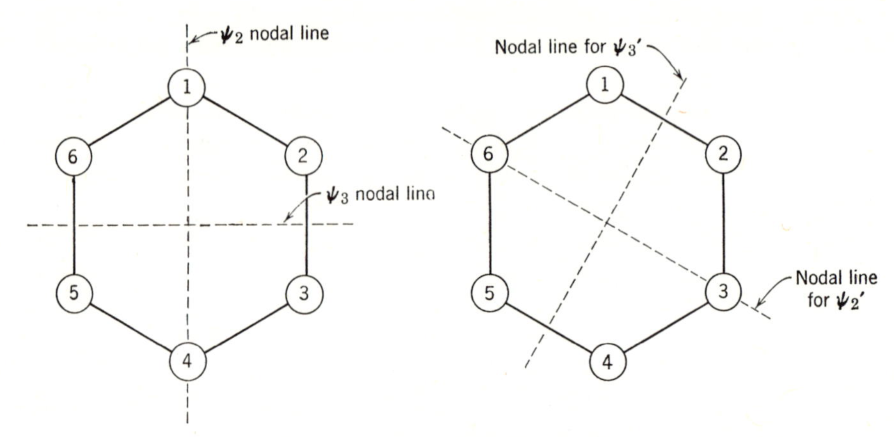

Figure 2.3 Nodal contours in benzene.

Figure 2.4 Alternative nodal contours in benzene.

for which the a_{ij} are the components of an orthogonal matrix are equally valid eigenvectors for this level, and each distinct such pair will rotate the nodal patterns of ψ_2 and ψ_3 through an arbitrary angle. This is a general phenomenon in quantum mechanics, and we shall see that alternative sets of degenerate eigenvectors can be transformed into each other by means of an orthogonal matrix.

Also important to us as scientists is the realization that quantum mechanics frequently confronts the investigator with undecidable propositions, such as the orientation of the nodes of the degenerate $\alpha + \beta$ molecular orbitals of benzene. It is a measure of the maturity of the student's grasp of our subject that he accept the existence of such undecidable propositions along with the realization that should he insist on imposing a decision on nature *philosophically*, none of the *experimental* properties (energy, electron density distribution, etc.) of the quantum mechanical system will be changed.

The indecidable propositions are thus removed from the realm of experimental confirmation into that of philosophic speculation, a region not forbidden to the scientist, but into which he enters stripped of any particular authority.

PROBLEMS

2.4 (1) For the benzene molecule, draw nodal lines through the Hückel diagram for the degenerate molecular orbitals ψ_4 and ψ_5 listed among equations 2.4 and indicate the algebraic sign of the wave function on each side of the nodes.

(2) Construct an alternative orthonormal pair ψ_4' and ψ_5' for the degenerate level $E_4 = E_5 = \alpha - \beta$ and show the new positions of the nodes.

2.5 Calculate the bond order of each bond in benzene in the ground state.

2.6 Hückel molecular orbital theory may be applied to conjugated structures containing a heteroatom. Thus formaldehyde $H_2C{=}O$ has a Hückel diagram analogous to ethylene except that the Hückel parameter α which is the (negative) energy of a valence electron in an isolated carbon atom must be lower than α for an isolated, more electronegative oxygen atom, and we *empirically* choose this lower energy to be $\alpha + \beta$. Thus the diagonal element H_{22} of the Hamiltonian matrix H is $H_{22} = \alpha + \beta$. The remaining matrix elements are unchanged.

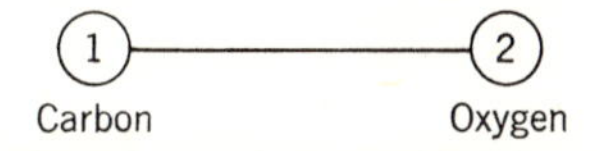

(1) Calculate the energy levels, ground state molecular energy, and Hückel molecular orbitals for formaldehyde.

(2) By comparing the total molecular energy with the energy of an electron in an isolated carbon atom plus the energy of another electron in an isolated oxygen atom, calculate the bond energy of the molecule in the ground state.

(3) Compute the electron density on the carbon and the oxygen atoms. The dipole moment of a molecule is a vector pointing from the negative toward the positive center of charge. What is the direction of the dipole moment of formaldehyde?

REFERENCES

1. A. S. Streitwieser, *Molecular Orbital Theory for Organic Chemists*, John Wiley and Sons, Inc., New York, 1961.
2. J. D. Roberts, *Notes on Molecular Orbital Calculations*, W. A. Benjamin, Inc., New York, 1962.
3. R. Daudel, R. Lefebvre, and C. Moser, *Quantum Chemistry, Methods and Applications*, Interscience, New York, 1959, Chapter IV.

Chapter 3

FUNCTION SPACES AND LINEAR OPERATORS

3.1 ANALOGY TO FINITE DIMENSIONAL VECTOR SPACES

Everything we have done so far has been within the framework of a finite dimensional vector space. The algebra we have developed is, however, abstractly analogous point by point to a much wider class of problems defined in an abstract space of infinitely many dimensions—a "function space." Thus suppose that we have two functions $\psi(x)$ and $\phi(x)$ which are finite and continuous over the range $a \leq x \leq b$. The integral

$$\int_a^b \psi(x)\phi(x)\, dx$$

is analogous to the formation of the *scalar product* of two vectors, where each "vector" ψ, ϕ has a continuum of components—one for each value of x in the range a, b. Similarly, two functions are said to be *orthogonal* in the range a, b if their scalar product vanishes

$$\int_a^b \psi(x)\phi(x)\, dx = 0 \tag{3.1}$$

and a function with the property

$$\int_a^b \psi^2(x)\, dx = 1 \tag{3.2}$$

is said to be *normalized* in the range a, b.

To illustrate equations 3.1 and 3.2 consider

$$\psi(x) = (\tfrac{3}{2})^{1/2}x; \qquad \phi(x) = (\tfrac{5}{8})^{1/2}(3x^2 - 1)$$

in the range $a = -1$; $b = 1$. The student may easily confirm that

$$\int_{-1}^{1} \psi^2(x)\, dx = \int_{-1}^{1} \phi^2(x)\, dx = 1; \qquad \int_{-1}^{1} \psi(x)\phi(x)\, dx = 0$$

We may also make linear transformations

$$\phi(x) = \int_a^b K(x, y)\psi(y)\, dy \tag{3.3}$$

where the function $K(x, y)$, corresponding to a matrix, is called the *kernel* of the equation. Thus if $K(x, y) = xy$, then using $\psi(x)$ above we have

$$\int_{-1}^{1} K(x, y)\psi(y)\, dy = (\tfrac{3}{2})^{1/2} \int_{-1}^{1} xy^2\, dy = (\tfrac{2}{3})^{1/2} x = (\tfrac{2}{3})\psi(x) \tag{3.4}$$

Analogs to the operations of matrix inversion and to the identity matrix I also exist, but we shall not need them. Important for our work, however, is the analog to the eigenvalue problem

$$\int_a^b K(x, y)\psi(y)\, dy = \lambda\psi(x)$$

which is illustrated by equation 3.4 with $\lambda = \frac{2}{3}$. Just as for matrices, such problems turn out to have solutions only if λ takes on one of a certain set of eigenvalues, and corresponding to each distinct such value λ_k are sets of *eigenfunctions* $\psi_k(x)$. The complication of degeneracy enters here also. For function spaces, however, there are complications that do not exist in finite dimensional vector spaces. There may be (and usually are) infinitely many eigenvalues and eigenfunctions, and the eigenvalues are not necessarily discrete; there may be a continuum of them.

It was convenient in Sections 1.13 to 1.15 to introduce a basis into a finite dimensional vector space in terms of which we could write expansions of the type

$$\mathbf{x} = c_1\mathbf{E}_1 + c_2\mathbf{E}_2 + \cdots + c_n\mathbf{E}_n$$

If the $\mathbf{E}_j$ are orthonormal, the expansion coefficients are

$$c_j = \mathbf{x} \cdot \mathbf{E}_j = \mathbf{x}^T\mathbf{E}_j$$

Similarly, in a function space it is possible to develop bases—sets of linearly independent functions in terms of which other "vectors" (finite, continuous functions) may be expanded:

$$\phi(x) = c_1\psi_1 + c_2\psi_2 + \cdots$$

where the sum may (and usually does) extend to infinity. If the ψ_j belong to a so called *orthonormal* set of functions, the expansion coefficients may be readily calculated. Thus for

$$\int_a^b \psi_i(x)\psi_j(x)\,dx = \delta_{ij}$$

c_j is the scalar product of ϕ with ψ,

$$c_j = \int_a^b \phi(x)\psi_j(x)\,dx$$

3.2 AN EXAMPLE

The set of functions $\psi_m = (2/\pi)^{1/2} \sin mx$; $m = 1, 2, \ldots$, is orthonormal over the range $0, \pi$. That is,

$$\begin{aligned}\int_0^\pi \psi_m(x)\psi_n(x)\,dx &= (2/\pi)\int_0^\pi \sin mx \sin nx\,dx \\ &= 0 \qquad \text{for} \quad m \neq n \\ &= 1 \qquad \text{for} \quad m = n\end{aligned}$$

Furthermore, the function $\phi(x) = 1$ is finite and continuous over the range $0, \pi$. Hence an expansion

$$1 = c_1\psi_1(x) + c_2\psi_2(x) + \cdots$$

should exist that will be valid for $0 < x < \pi$. The coefficients are

$$\begin{aligned}c_m &= \int_0^\pi (1)\psi_m(x)\,dx = (2/\pi)^{1/2}\int_0^\pi \sin mx\,dx \\ &= 0 \qquad \text{if} \quad m \text{ is even} \\ &= (2/\pi)^{1/2}(2/m) \qquad \text{if} \quad m \text{ is odd.}\end{aligned}$$

Hence

$$1 = (4/\pi)\{\sin x + (\tfrac{1}{3}) \sin 3x + (\tfrac{1}{5}) \sin 5x + \cdots\}$$

PROBLEM

3.1 Using the basis set $\psi_m(x) = (2/\pi)^{1/2} \sin mx$; $m = 1, 2, \ldots$, calculate the coefficients c_m in the expansion

$$x = \sum_{m=1}^{\infty} c_m\psi_m(x)$$

valid for $0 \leq x < \pi$ and write out explicitly the first several terms.

3.3 COMPLEX FUNCTION SPACES

In Section 1.22 we found that the eigenvectors of a real, symmetric matrix could be arranged into an orthonormal set that was a suitable basis in the vector space of n dimensions. Similarly, the eigenfunctions of a *symmetric kernel* $K(x, y) = K(y, x)$ can be arranged into orthonormal sets of functions, and these functions then form a suitable basis in function space. A further generalization of function spaces to include complex numbers leads to a definition of the scalar product of two functions as

$$\int_a^b \phi^*(x)\psi(x)\,dx \tag{3.5}$$

and of the norm of a function ψ as the positive square root of

$$\int_a^b \psi^*(x)\psi(x)\,dx \tag{3.6}$$

In both of these integrals the asterisk implies complex conjugation.

Examples of equations 3.5 and 3.6 are

$$\int_0^{2\pi} (e^{2ix})^*(e^{ix})\,dx = \int_0^{2\pi} e^{-2ix}e^{ix}\,dx = \int_0^{2\pi} e^{-ix}\,dx = 0$$

$$\int_0^{2\pi} (e^{ix})^*(e^{ix})\,dx = \int_0^{2\pi} e^{-ix}e^{ix}\,dx = \int_0^{2\pi} dx = 2\pi$$

in which $i = \sqrt{-1}$.

3.4 INTEGRAL AND DIFFERENTIAL EQUATIONS

Our discussion of linear transformations in function space has so far been based on the integral relationship

$$\phi(x) = \int_a^b K(x, y)\psi(y)\,dy$$

and in particular I have stated the eigenvalue problem as

$$\int_a^b K(x, y)\psi(y)\,dy = \lambda\psi(x)$$

The last formula is a particular type of integral equation which requires us to find λ and ψ if we are given K. The statement of the problem in this form is suggestive because of its analogy to matrix algebra. In quantum mechanics, however, it is not customary to state eigenvalue problems as integral equations. They are almost without exception stated as differential equations. I shall joyously pass over the detailed transformation of an integral equation into a differential equation and be content to state that an integral equation can be made equivalent to a differential equation *plus boundary conditions.* This last qualification is very important, for the integral equation contains within itself all of the boundary conditions of a particular problem, and no extra conditions of this sort need be imposed upon a solution to an integral equation. Once the integral equation has been rewritten as a differential equation, however, it turns out that the number of solutions to the differential equation is greater than those that satisfy the integral equation, so that boundary conditions must be imposed in order to eliminate the unwanted solutions. That this should be so is not unreasonable when the student recalls from his study of ordinary differential equations that arbitrary constants always appear in the general solution and that these arbitrary constants must then be determined from exterior constraints imposed upon the general solution.

3.5 AN EXAMPLE

A differential equation with boundary conditions that leads to a problem of the eigenvalue type is

$$\frac{d^2\psi}{dx^2} = -\lambda\psi \tag{3.7}$$

where we require $\psi(0) = \psi(\pi) = 0$ as boundary conditions. The general solution of the differential equation is

$$\psi = A \sin \sqrt{\lambda}\, x + B \cos \sqrt{\lambda}\, x$$

with A and B arbitrary constants. At $x = 0$, $\psi = B$, so that we must put $B = 0$,

$$\psi = A \sin \sqrt{\lambda}\, x$$

Now, for arbitrary values of λ it is impossible to satisfy the second boundary condition. That this should be so may be inferred from a study of Figure 3.1 in which $\sin \sqrt{\lambda}\, x$ is plotted against x for several values of λ. As λ increases from 0, the first positive root of $\sin \sqrt{\lambda}\, x$ exceeds $x = \pi$. For $\lambda = 1$ the first root falls on $x = \pi$, but with increasing λ, $\sin \sqrt{\lambda}\, x$ is not again zero at $x = \pi$ until $\lambda = 4$. There are, in short, only a set of discrete choices $\sqrt{\lambda} = n$, an integer, for which the second boundary condition can be satisfied. We

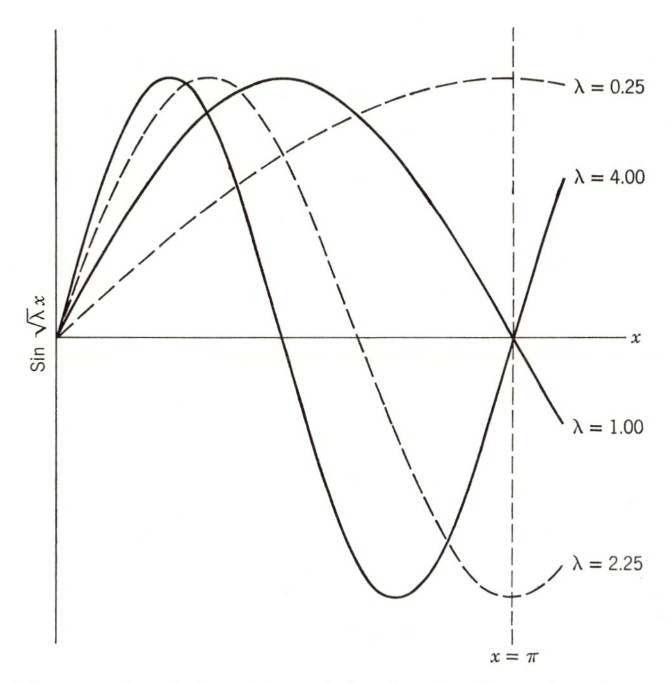

Figure 3.1 The generation of eigenvalues and eigenfunctions from a boundary condition.

are thus led to a spectrum of eigenvalues $\lambda = n^2$; $n = 1, 2, \ldots$ and to a set of associated eigenfunctions $\psi_n = A \sin nx$. We have already observed in Section 3.2 that these eigenfunctions are orthogonal, and if we set $A = (2/\pi)^{1/2}$, they will also be normalized.

For future reference let the student note that equation 3.7 is equally well satisfied by

$$\psi = C \exp(i\sqrt{\lambda}\, x) + D \exp(-i\sqrt{\lambda}\, x)$$

with arbitrary constants C and D. For ψ to vanish at $x = 0$, we need $C = -D$, or

$$\psi = C\{\exp(i\sqrt{\lambda}\, x) - \exp(-i\sqrt{\lambda}\, x)\} = 2iC \sin \sqrt{\lambda}\, x$$

which is identical with our results above if $\lambda = n^2$ and $C = (1/2i)(2/\pi)^{1/2}$.

3.6 THE ONE DIMENSIONAL HARMONIC OSCILLATOR

Let us apply these ideas to a quantum mechanical problem, cheerfully disregarding the postulential basis that leads to the methods employed. In classical mechanical terminology we are asked to describe the motion of a particle of mass m constrained to move along the x axis under the influence of a spring binding it to the origin $x = 0$. The force exerted on the particle is given by Hooke's law as proportional to the displacement x: force $= -kx$; but the same information is conveyed in classical mechanics by stating that the potential energy of the particle is $V = \frac{1}{2}kx^2$. The solution of Newton's equations of motion for this problem is

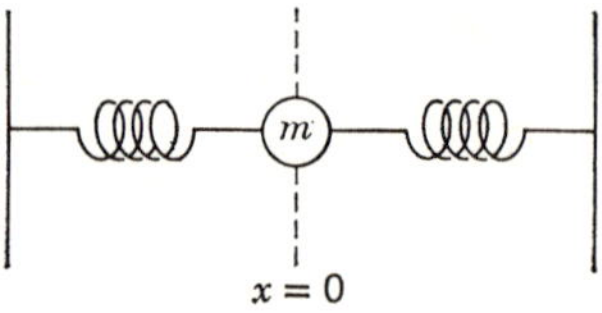

$$x = A \cos 2\pi\nu(t - \phi) \tag{3.8}$$

in which $\nu = \dfrac{1}{2\pi}(k/m)^{1/2}$ is the frequency of the motion in oscillations per second, A is the amplitude of the motion $-A \leq x \leq A$, and ϕ is the "phase angle." Both A and ϕ are integration constants, so that in any particular case they must be determined through experiment before this formula can be used in numerical calculations. Note that a solution to a classical mechanical problem gives the position coordinate x as a function of time.

The position x of a particle is only one of the important dynamical *observables* of classical mechanics. Another important observable is the momentum p associated with motion along the x axis. (An observable is a quantity that can be measured, i.e., "observed.") The momentum in this case may be adequately defined as $p =$ mass $\times$ velocity $= m(dx/dt)$, or

$$p = -2\pi m\nu A \sin 2\pi\nu(t - \phi) \tag{3.9}$$

Now, if at $t = 0$ we can simultaneously measure by an experiment the values of both x and p:

$$x(0) = A \cos 2\pi\nu\phi$$

$$p(0) = 2\pi m\nu A \sin 2\pi\nu\phi$$

then both A and ϕ can be calculated from these data and the motion is thenceforth completely predictable for all time.

Another important observable of classical mechanics is the total energy, kinetic + potential, of the system. In the present case, the total energy is

$$E = \frac{1}{2}m\left(\frac{dx}{dt}\right)^2 + \frac{1}{2}kx^2$$

$$= \frac{p^2}{2m} + \frac{1}{2}kx^2$$

whence it follows from (3.8) and (3.9) that

$$E = 2m\pi^2\nu^2A^2 = \text{independent of time}$$

Thus as the amplitude A is changed, E changes continuously, so that, depending on the initial conditions, all positive values of E are permitted.

3.7 QUANTUM MECHANICAL TREATMENT OF THE ONE DIMENSIONAL, HARMONIC OSCILLATOR

In this book we are exclusively interested in problems that in classical mechanics would involve "bound" particles—particles that execute periodic or quasi periodic motion in a restricted region of space. Examples are the planetary motion of satellites around the sun, the wanderings of a particle inside a box, and the spinning of a sphere around an axis. A counterexample would be the fast motion of a projectile passing through a medium and not returning to it. A distinguishing feature of bound state problems in quantum mechanics is that their measurable properties do not involve the time. This approach is radically different from classical dynamical problems, whose solutions, as we have just observed, involve writing down the position coordinates as functions of time. The periodic or quasi periodic trajectories of a classical bound particle become blurred into a sort of timeless continuum in the quantum mechanical treatment.

In quantum mechanics we are, in fact, not asked to find x as a function of t. Rather, we are to find a function $\psi(x)$ from which all experimentally observable information about the oscillator can be calculated. Acceptable functions $\psi(x)$ must be eigenfunctions of a certain differential operator (the Hamiltonian operator H), and they must satisfy certain boundary conditions. These last are of major importance, for they select from the totality of eigenfunctions of H a set that for bound particles always turns out to be discrete as well as a suitable basis for the expansion of arbitrary, finite, continuous functions as linear combinations of eigenfunctions. In the present case, the Hamiltonian operator for the one dimensional, harmonic oscillator is

$$H = -\frac{\hbar^2}{2m}\frac{d^2}{dx^2} + \tfrac{1}{2}kx^2$$

in which $\hbar = h/2\pi$ with $h =$ the quantum constant of Planck

$$h = 6.626 \times 10^{-27} \text{ erg sec}$$

$$\hbar = 1.055 \times 10^{-27} \text{ erg sec}$$

Our eigenvalue problem may be stated as

$$H\psi = E\psi$$

with E the as yet undetermined eigenvalue, so that

$$-\frac{\hbar^2}{2m}\frac{d^2\psi}{dx^2} + \tfrac{1}{2}kx^2\psi = E\psi \tag{3.10}$$

This is a differential equation (Schrödinger's equation) for the determination of ψ, but only those solutions of (3.10) are acceptable that are finite and continuous over the range $-\infty \leq x \leq \infty$ and are normalized in the Hermitian sense

$$\int_{-\infty}^{\infty} \psi^*\psi \, dx = 1$$

Upon investigation[1-3] it turns out that for arbitrary values of E the solutions of the differential equation usually become arbitrarily large as $x \to \infty$. Such a function cannot be normalized, so that we exclude it from the class of acceptable eigenfunctions. It follows that unless we are willing to tamper with E, there are no solutions to our overall eigenvalue problem. Only if E is restricted to the discrete set of values

$$\begin{aligned} E_n &= \hbar\left(\frac{k}{m}\right)^{1/2}(n + \tfrac{1}{2}) \\ &= (n + \tfrac{1}{2})h\nu; \qquad n = 0, 1, 2, \ldots \end{aligned} \tag{3.11}$$

will acceptable eigenfunctions exist. This is, of course, exactly analogous to the situation in a finite dimensional vector space and to that outlined in Section 3.5: eigenvectors do not exist for arbitrary eigenvalues λ. They exist only for a certain discrete set.

Corresponding to the eigenvalues E_n of equation 3.11 are normalized eigenfunctions

$$\psi_n(\xi) = [(\pi x_0)^{1/2} 2^n n!]^{-1/2} H_n(\xi) \exp(-\tfrac{1}{2}\xi^2) \tag{3.12}$$

where $\xi = x/x_0$ and $x_0 = (\hbar/\sqrt{mk})^{1/2} = \left(\frac{1}{2\pi}\right)(h/\nu m)^{1/2}$ is a fundamental unit of length for the oscillator. The $H_n(\xi)$ are known in mathematics as the Hermite polynomials. Their first several members are

$$H_0(\xi) = 1; \qquad H_1(\xi) = 2\xi; \qquad H_2(\xi) = 4\xi^2 - 2; \ldots$$

Plots of the first three eigenfunctions (3.12) are sketched in Figure 3.2.

Now that we have the eigenfunctions and eigenvalues of the Hamiltonian operator for the one dimensional, harmonic oscillator we are left with the job of interpreting them, and here the postulates of quantum mechanics

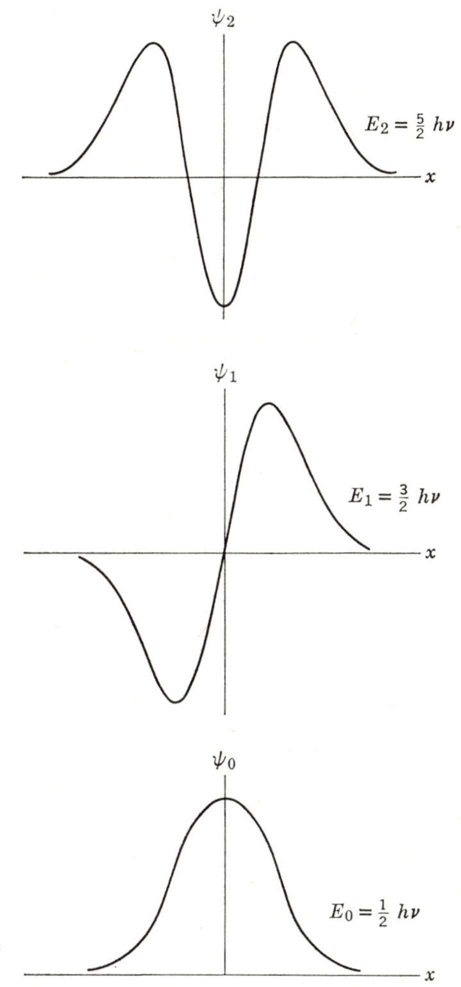

Figure 3.2 Wave functions for the linear harmonic oscillator.

come to our aid. These postulates state that for every dynamical observable of classical mechanics (position, momentum, energy, etc.) there corresponds in quantum mechanics an operator from which numerical information is calculated via an integration:

$$\text{expected observable value} = \int \psi^*(\text{operator})\psi \, dx$$

Note the analogy to matrix algebra: an operator is there a linear transformation represented by a matrix, say K. Then if the state of a mechanical system is represented by a vector ψ, the expectation of the observable scalar quantity k is the scalar product of ψ into the transformed vector $K\psi$:

$$k = \psi^T K \psi$$

Geometrically, for ψ a unit vector and not an eigenvector of K, $K\psi$ has a

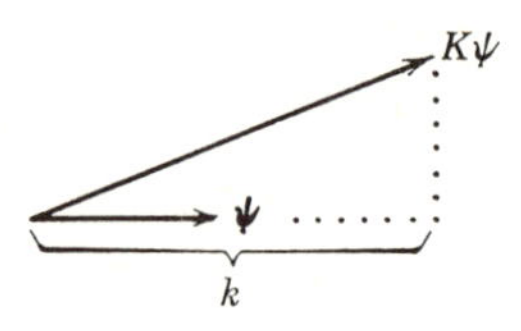

direction different from ψ. The expectation $k = \psi^T K \psi$ is the component of $K\psi$ in the direction of ψ. If ψ happens to be an eigenvector of K, then

$$K\psi = \lambda\psi$$

and the component of $K\psi$ in the direction of ψ,

$$k = \psi^T K \psi = \psi^T \lambda \psi = \lambda \psi^T \psi = \lambda$$

is just the eigenvalue λ.

Now, the operator corresponding to the total energy in quantum mechanics is the Hamiltonian operator. Observable values of the total energy are therefore

$$\int \psi_n^* H \psi_n \, dx = \int \psi_n^* E_n \psi_n \, dx = E_n \int \psi_n^* \psi_n \, dx = E_n$$

and the eigenvalues of the Hamiltonian operator are thus the allowed observable values of the total energy. For the harmonic oscillator (and it turns out always for problems involving bound particles) these energy eigenvalues are discrete. They are called *energy levels*.

A more detailed interpretation of the state "vector" or eigenfunction ψ is given in the next chapter.

PROBLEMS

3.2 Show that $\psi_0(x) = (\pi x_0^2)^{-1/4} \exp[-\frac{1}{2}(x/x_0)^2]$; $E_0 = \frac{1}{2}h\nu$ satisfies Schrödinger's equation 3.10 for the linear, harmonic oscillator.

3.3 Given $\psi_1(x) = (2^{1/2}\pi^{-1/4}x_0^{-3/2})x \exp[-\frac{1}{2}(x/x_0)^2]$ and ψ_0 as in Problem 3.2, calculate the integrals

$$\int_{-\infty}^{\infty} {\psi_0}^2\,dx; \qquad \int_{-\infty}^{\infty} {\psi_1}^2\,dx; \qquad \int_{-\infty}^{\infty} \psi_0\psi_1\,dx$$

3.4 For the one dimensional, harmonic oscillator the quantum mechanical operator $P = (\hbar/i)(d/dx)$ corresponds to classical linear momentum $p = m(dx/dt)$ along the x axis.

(1) Calculate the expectation of the momentum in the ground state:

$$\langle p \rangle = \int_{-\infty}^{\infty} \psi_0^* P \psi_0\,dx$$

(2) Calculate the expectation of the square of the momentum in the ground state

$$\langle p^2 \rangle = \int_{-\infty}^{\infty} \psi_0^* P^2 \psi_0\,dx$$

3.5 For the one dimensional harmonic oscillator, the quantum mechanical operator for the position coordinate is x and the quantum mechanical operator for the momentum is $P = (\hbar/i)(d/dx)$. Show that x and P do not commute, $xP \neq Px$, that is, for *any* nonzero function $f(x)$

$$xPf(x) \neq Pxf(x)$$

3.6 For the one dimensional harmonic oscillator in the ground state calculate the expectation $\langle x \rangle$ of x and the expectation $\langle x^2 \rangle$ of x^2. The statistical standard deviation

$$\delta x = (\langle x^2 \rangle - \langle x \rangle^2)^{1/2}$$

is a measure of the experimental uncertainty in any measurement of the position of the particle. Similarly, the quantity calculated from Problem 3.4,

$$\delta p = (\langle p^2 \rangle - \langle p \rangle^2)^{1/2}$$

is a measure of the experimental uncertainty in any determination of the momentum of the particle.

Calculate the product $\delta x\,\delta p$ and show that it is of the order of magnitude of $\hbar$. The result is a special case of the Heisenberg Uncertainty Principle.

3.7 A typical, macroscopic, linear harmonic oscillator of the type that could be constructed in a machine shop has a Hooke's law constant of $k = 10{,}000$ dyne/cm and a particle mass of $m = 1$ g.

(1) Using the physical constants listed below, calculate an approximate quantum number n for this oscillator when its energy is equal to

the thermal energy $E = kT$ at room temperature $T = 300°K$. Here k is the Boltzmann constant.

(2) Repeat this calculation when the particle of mass 1 g is replaced by a proton.

Planck's constant $h = 6.63 \times 10^{-27}$ erg sec.
Boltzmann's constant $k = 1.38 \times 10^{-16}$ erg deg^{-1}.
mass of the proton $= 1.67 \times 10^{-24}$ g.

REFERENCES

1. L. Pauling and E. B. Wilson, Jr., *Introduction to Quantum Mechanics*, McGraw-Hill, Inc., New York, 1935, pp. 67–73.
2. H. Eyring, J. Walter, and G. E. Kimball, *Quantum Chemistry*, John Wiley and Sons, Inc., New York, 1944, pp 75–79.
3. W. Kauzmann, *Quantum Chemistry*, Academic Press, Inc., New York, 1957, pp. 201–207.

Chapter 4

QUANTUM MECHANICS FOR BOUND STATES

In this chapter we introduce the postulates of quantum mechanics needed for the study of bound state problems. The reader will recall that such problems when examined classically involve motion that is restricted by potential fields to a particular region of space and for which the motion is thus of a periodic or quasi periodic nature.

4.1 CONFIGURATION SPACE

In classical mechanics the motion of a set of particles takes place in ordinary three dimensional space, and to locate each particle with respect to some set of coordinate axes fixed in the laboratory, a triple of numbers (say x, y, z) is assigned to each particle. When the position of all particles is known, this is equivalent to saying that we know the numerical values of all the coordinates of all the particles, so that we know the *configuration* of the system. This list of information may equivalently be interpreted by writing the coordinates of the first particle x_1, y_1, z_1 together with those of the second x_2, y_2, $z_2, \ldots$, together with those of the Nth, x_N, y_N, z_N in a single row $(x_1, y_1, z_1, x_2, y_2, z_2, \ldots, x_N, y_N, z_N)$, so that the configuration of the system is equivalent to knowing a point in a *configuration space* of $3N$ dimensions. To every such point in configuration space corresponds some different configuration of all the N particles.

4.2 CONJUGATE MOMENTA

Associated with each coordinate of a classical mechanical particle is a momentum. Thus for the x coordinate of the first particle there is associated a momentum p_{x1}, with the z coordinate of the fourth particle a momentum p_{z4}, and so on. The number of associated or *conjugate* momenta is hence equal to the number of coordinates, and for each point in the configuration space of the system is associated a list of $3N$ conjugate momenta. When the coordinates used to describe the system are Cartesian coordinates, the conjugate momenta are known as linear momenta and are related to the velocities of the particles by $p_{x1} = m_1(dx_1/dt)$; $p_{y1} = m_1(dy_1/dt)$; . . . ; $p_{zN} = m_N(dz_N/dt)$, in which m_j is the mass of the jth particle. More often than not, however, Cartesian coordinates are not the most convenient ones with which to describe the motion of a classical mechanical system. When other coordinate systems are used (such as spherical polar coordinates r, θ, φ), the dimension of the configuration space does not change (it is still $3N$), but the representative point would be written $(r_1, \theta_1, \varphi_1, r_2, \theta_2, \varphi_2, \ldots, r_N, \theta_N, \varphi_N)$. The list of conjugate momenta still contains $3N$ entries, but they are less simply related to the time derivatives dr/dt, $d\theta/dt$, $d\varphi/dt$ of the particle coordinates. Rules for the construction of the conjugate momenta for coordinate systems other than Cartesian are given in standard texts,[1,2] but we need not be concerned with them here. I shall only remark that momenta conjugate to the angular coordinates θ, φ are known as *angular momenta*.

4.3 EQUATIONS OF MOTION

A problem in classical mechanics may be said to have been solved if we know all the coordinates of all the particles as functions of the time. According to the formulation of classical mechanics in the hands of the Irish mathematician W. R. Hamilton in the first half of the nineteenth century, such an explicit solution may be obtained by the integration of certain differential equations of motion, provided that at some specific instant of time (say $t = 0$) all of the coordinates and all of their conjugate momenta are known exactly through an experimental measurement. Once this experimental information is available and the equations of motion have been integrated, the coordinates are predicted to change with time in such a way that the representative point of the system traces out a certain trajectory through its configuration space. The student should distinguish carefully between the motion of this mathematical point through a configuration space of $3N$ dimensions and the physically observable motion of N discrete particles through the three dimensional space of the laboratory.

Classical mechanical systems that execute periodic or quasi periodic motion have representative points moving through a restricted region of configuration space, so that after the passage of a long period of time the representative point may be said to have swept out a certain localized volume of configuration space. Such problems correspond in quantum mechanics to bound state problems.

4.4 AN HYPOTHESIS OF IGNORANCE

The idea of the representative point of a classical particle or of a system of particles sweeping out a particular region of configuration space by making repeated trajectories through it is a useful one, for it enables us to conceive of the idea of a probability density defined in the space in such a way that the probability of observing the representative point in a small volume $d\tau = dx_1\, dy_1\, dz_1\, dx_2\, dy_2\, dz_2 \cdots dx_N\, dy_N\, dz_N$ of configuration space is proportional to the relative amount of time the representative point spends there. (This means the probability of observing a configuration such that the coordinates of the first particle lie in a small volume $dx_1\, dy_1\, dz_1$ centered on x_1, y_1, z_1 in physical or laboratory space while simultaneously the coordinates of the second lie in a small volume $dx_2\, dy_2\, dz_2$ centered on x_2, y_2, z_2, and so on.) Such statistical considerations are not intrinsic to classical mechanics, and they play no role in the integration of Newton's (or, what is equivalent, Hamilton's) equations of motion. They are, however, essential to quantum mechanics; and a cardinal philosophical position of quantum mechanics is that we may know only statistical details of the trajectory of a representative point through its configuration space. It is for this reason that time does not enter into bound state problems, and one interpretation of bound state quantum mechanics is that it describes the time averaged trajectories of classical particles. I hasten to add, however, that while the regions of configuration space filled with classical trajectories for a given system correspond roughly to those described statistically by quantum mechanics, the details of the statistical distribution as calculated by the two disciplines are not the same.

These remarks are well illustrated by the one dimensional, harmonic oscillator of Section 3.6. The classical motion is described by equation 3.8:

$$x = A \cos 2\pi\nu(t - \phi)$$

Because the motion is strictly periodic, we do not need to consider long periods of time, but only the time $\frac{1}{2}\nu^{-1}$ required for a half cycle.

The time dt required for the oscillator to make a single passage through the region dx centered on x is

$$dt = \left|\frac{dx}{dt}\right|^{-1} dx$$

$$= |A2\pi\nu \sin 2\pi\nu(t - \phi)|^{-1}\, dx$$

$$= (A2\pi\nu)^{-1}\left(1 - \frac{x^2}{A^2}\right)^{-\frac{1}{2}} dx$$

The average time spent in the neighborhood of x is then

$$\frac{dt}{\frac{1}{2}\nu^{-1}} = \frac{1}{A\pi}\left(1 - \frac{x^2}{A^2}\right)^{-\frac{1}{2}} dx$$

and we shall identify this with a probability distribution function

$$P(x) = \frac{1}{A\pi}\left(1 - \frac{x^2}{A^2}\right)^{-\frac{1}{2}}$$

such that the probability of observing the oscillating particle at x to within an accuracy dx is $P(x)\, dx$. The student should check the reasonableness of this statement by calculating

$$\int_{-A}^{A} P(x)\, dx = 1$$

meaning that the total probability of finding the particle somewhere between $x = -A$ and $x = A$ is unity.

$P(x)$ is plotted in Figure 4.1, whence it appears that the classical particle spends most of its time near the extremes of the motion where its speed $|dx/dt|$ is least.

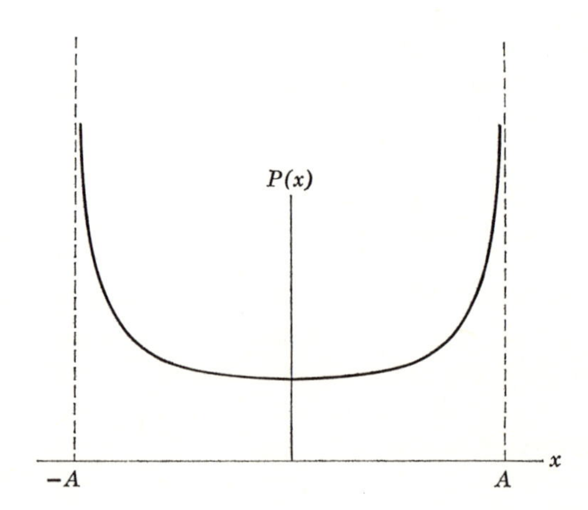

Figure 4.1 Probability distribution in configuration space for a classical oscillator.

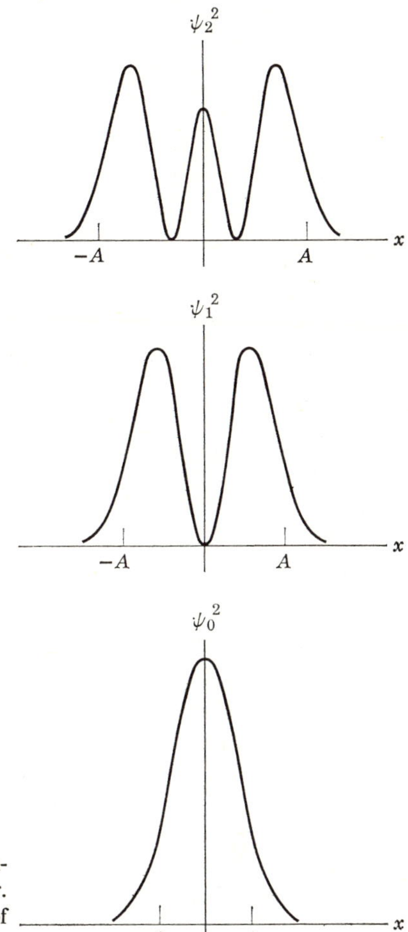

Figure 4.2 Probability distributions in configuration space for a quantum oscillator. A_n is the amplitude of a classical oscillator of energy $E_n = (n + \frac{1}{2})h\nu$.

The corresponding quantum mechanical probability distribution functions are the squares of the wave functions ψ that were sketched in Figure 3.2. A comparison with $P(x)$ of ψ_0^2, ψ_1^2, ψ_2^2 drawn in Figure 4.2 shows striking differences, for not only do the quantum distributions exhibit nodes where the probability of observing the particle is zero, but there is a nonvanishing probability that the quantum particle will be observed beyond the limits of the classical motion at $x = \pm A$.

Despite the differences between Figures 4.1 and 4.2, however, Figure 4.3 illustrates the interesting fact[3,4] that as n becomes large, a curve drawn through all of the inflection points of ψ_n^2 begins to look more and more like the classical profile sketched in Figure 4.1.

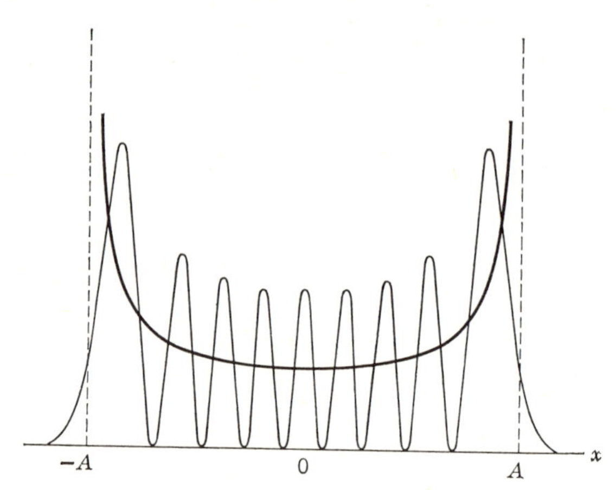

Figure 4.3 Comparison of ψ_8^2 for the quantum oscillator with the probability distribution function $P(x)$ of a classical oscillator having the same energy. A is the classical amplitude.

4.5 OBSERVABLES

An observable in classical mechanics is a quantity that can be measured experimentally. Thus the three position coordinates q_1, q_2, q_3 and conjugate momenta p_1, p_2, p_3 of a classical particle are capable of being observed as are also its kinetic and potential energies. Other observables exist, but we need not make a complete list at this point. The purely kinematic observables of position and momentum are the basic ones for those classical mechanical problems whose quantum mechanical analogs we shall be studying. For according to Hamilton's formulation of classical mechanics, from the three position coordinates and their three conjugate momenta the kinetic and the potential energies of a particle may be calculated, and we may facetiously acknowledge this fact by stating that if we know our p's and q's, then all other classical observables are known. In generalized coordinate systems (not necessarily Cartesian) the kinetic energy K of a particle is a function of both its position and its momentum, but in this book we are concerned only with potential energies which are functions of position only. These remarks have to be extended to include the position coordinates and conjugate momenta of all particles for a system of particles; for systems of charged particles they have to be modified to include certain electromagnetic effects. Parenthetically it may be noted that the properties of charge and mass are extraneous both to classical mechanics and to quantum mechanics in the form in which we present them, where they enter only as parameters. Their proper inclusion as observables in a more general field theory remains an unsolved problem.

4.6 POSTULATES

Quantum mechanics begins by defining a function (the "wave" function) in the configuration space of a classical mechanical system $\psi = \psi(x_1, y_1, z_1, \ldots, x_N, y_N, z_N)$. This function has the following properties:

1. It is finite, single valued, and continuous throughout the configuration space.

2. The probability of observing the classical particles in a configuration whose representative point lies in a small volume $d\tau$ of configuration space centered on $(x_1, \ldots, z_N)$ is $\psi^*\psi\, d\tau$, where the asterisk (*) implies complex conjugation of ψ in the case that ψ is a complex number. If ψ is real, then $\psi^*\psi$ reduces to ψ^2. Because the total probability of observing the representative point of the system somewhere in its configuration space is unity, ψ is normalized according to the rule

$$\int \psi^*\psi\, d\tau = \underbrace{\int_{-\infty}^{\infty} \cdots \int_{-\infty}^{\infty}}_{3N \text{ fold multiple integral}} \psi^*(x_1, \ldots, z_N)\psi(x_1, \ldots, z_N)\, dx_1 \cdots dz_N = 1$$

3. To every observable of classical mechanics there corresponds in quantum mechanics an operator that acts on the function ψ to produce another function. If the observable is some position coordinate q (i.e., an x, y, z, or an r, θ, φ), the operation is multiplication of ψ by q. If the observable is a function V of the position coordinates (such as the potential energy $V(x_1, \ldots, z_N)$ of the system), the operation is multiplication of ψ by V. If the observable is a momentum conjugate to q, then for a Cartesian coordinate q the operator is

$$\frac{\hbar}{i}\frac{\partial}{\partial q}$$

where $\hbar$ = Planck's constant divided by 2π and $i = \sqrt{-1}$. Thus, for example, the operator corresponding to p_{x1} is

$$\frac{\hbar}{i}\frac{\partial}{\partial x_1}$$

meaning that ψ is to be differentiated partially with respect to x_1 followed by multiplication with $\hbar/i$. If the observable is a function $K(p_1, p_2, \ldots)$ of various Cartesian conjugate momenta, the operator that acts on ψ is

$$K\left[\frac{\hbar}{i}\frac{\partial}{\partial q_1}, \quad \frac{\hbar}{i}\frac{\partial}{\partial q_2}, \quad \ldots\right]$$

4. Corresponding to some classical observable s we construct from the above rules an operator S. Then the mean value of S that may be observed in an experiment is calculated from the wave function via an integration,

$$\langle s \rangle = \int \psi^* S \psi \, d\tau$$

where the $3N$ fold integration is carried out over the entirety of configuration space. A case of special importance arises if ψ is an eigenfunction of S,

$$S\psi = \lambda\psi$$

for then

$$\langle s \rangle = \int \psi^* S \psi \, d\tau = \lambda \int \psi^* \psi \, d\tau = \lambda$$

so that the observable s has precisely the value λ. If ψ is not an eigenfunction of S, then the result of any measurement of s will still be some eigenvalue of S, but which eigenvalue in any particular experiment is not predictable; for our calculation yields only the statistical average $\langle s \rangle$ of a large number of experiments measuring s. For ψ an eigenfunction of S, the result of all experiments measuring s will always be λ.

5. Permitted wave functions for the quantum mechanical system are eigenfunctions of the Hamiltonian operator H:

$$H\psi = E\psi$$

This is the Schrödinger equation. The Hamiltonian operator in quantum mechanics corresponds to the classical observable of the total energy E in classical mechanics and is defined to be the sum of the kinetic and potential energy operators:

$$H = K + V$$

According to postulate 4, the eigenvalue E associated with the eigenfunction ψ is the total energy of the system when it is in the state described by ψ. We are thus always permitted to know precisely the total energy of a quantum mechanical system in a bound state, but the student should realize that ψ, while it is always an eigenfunction of H, may for certain systems also be an eigenfunction of other operators, and that for these systems it will be possible to know precisely observables other than E.

4.7 THE HARMONIC OSCILLATOR

Let us now return to a problem considered briefly in Section 3.7. There I wrote down the Schrödinger equation for a one dimensional harmonic oscillator and discussed some aspects of its solution, showing in particular how the differential equation and its boundary conditions led to the identification of a discrete set of permitted eigenfunctions for the oscillator together with an associated discrete set of eigenvalues or energy levels. It remains for us here to show how the Schrödinger equation can be constructed from the postulates we have just discussed.

The first step in the exact solution of any quantum mechanical problem is the construction of the Hamiltonian operator. The construction begins on a strictly classical basis. Write down the potential and kinetic energies of the classical system as a function of the Cartesian position coordinates and their conjugate momenta. Then substitute for the conjugate momenta $p_1, p_2, \ldots$ occurring in the classical kinetic energy the operators $(\hbar/i)(\partial/\partial q_1)$, $(\hbar/i)(\partial/\partial q_2), \ldots$. The result is the kinetic energy operator K, a differential operator. The potential energy operator V is identical with the classical potential energy function $V(q_1, q_2, \ldots)$. The final Hamiltonian operator is then the sum

$$H = K + V$$

and the Schrödinger equation is

$$H\psi = E\psi$$

with E the as yet undetermined eigenvalue.

For the one dimensional harmonic oscillator, a single position coordinate x is sufficient to locate the particle, so that the configuration space of the oscillator is one dimensional. There is correspondingly a single conjugate momentum p, and from Section 3.6 the classical potential energy is $V = \frac{1}{2}kx^2$ and the classical kinetic energy $K = p^2/2m$. The potential energy operator in quantum mechanics is identical with the classical V. To form the kinetic energy operator, we substitute for p in the classical expression the differential operator $(\hbar/i)(d/dx)$. Thus

$$K = \frac{[(\hbar/i)(d/dx)]^2}{2m} = -\frac{\hbar^2}{2m}\frac{d^2}{dx^2}$$

Finally, the Hamiltonian operator is the sum of the kinetic and potential energy operators

$$H = K + V = -\frac{\hbar^2}{2m}\frac{d^2}{dx^2} + \frac{1}{2}kx^2$$

Setting up the Schrödinger equation $H\psi = E\psi$, we have the differential equation 3.10 whose solution was studied in Section 3.7. The replacement of partial differentiation with respect to x by ordinary differentiation is justified in this instance only because of the one dimensional character of the configuration space. More complex problems invariably lead to problems in partial differential equations.

4.8 HERMITIAN OPERATORS

After the elaborate preparation we have had with symmetric, real matrices and Hermitian matrices, it will surprise no one to discover that the differential operators with which we have to deal in quantum mechanics have sharply analogous properties. The only difference between the two, in fact, is that while matrices are designed to operate on finite dimensional vector spaces, the differential operators of quantum mechanics operate on function spaces. The "state vectors" or "wave functions" with which we shall have to deal have properties analogous to finite dimensional vectors, but the computational details are more complicated. Thus the scalar product of two wave functions ϕ and ψ is defined to be

$$\int \phi^* \psi \, d\tau = \underbrace{\int_{-\infty}^{\infty} \cdots \int_{-\infty}^{\infty}}_{3N \text{ fold multiple integral}} \phi^*(x_1, \ldots, z_N)\psi(x_1, \ldots, z_N)\, dx_1 \cdots dz_N \tag{4.1}$$

If this scalar product vanishes, ϕ and ψ are said to be orthogonal.

All of the operators that will occur in our work are Hermitian. For a formal treatment of Hermitian operators in function spaces, the reader is referred elsewhere.[5-8] I shall be content to state the following theorems by analogy with what we have already written down for real, symmetric matrices.

1. Eigenvalues of Hermitian operators are real numbers.
2. Eigenfunctions belonging to different eigenvalues of an Hermitian operator are orthogonal under the scalar product rule (4.1). If an eigenvalue is n fold degenerate, a set of n linearly independent eigenfunctions having this eigenvalue can always be found, and they can by convention always be chosen orthogonal. Infinitely many such sets of n fold orthogonal eigenfunctions exist, and no one such set is to be preferred *ab initio* over any other.
3. Hermitian operators that commute possess eigenfunctions in common.

4.9 THE TWO DIMENSIONAL, ISOTROPIC, HARMONIC OSCILLATOR

In this section I take up a more complicated Schrödinger equation than those considered heretofore. This book does not constitute a treatise on differential equations, and as with the one dimensional harmonic oscillator, fine points concerning the solution of differential equations will be found elsewhere.[9,10] The two dimensional, isotropic oscillator will, however, serve to illustrate the separation method of solving partial differential equations, the use of a non-Cartesian coordinate system (plane polar coordinates), and the significance of commuting operators in quantum mechanics. We shall also derive some useful analogies to the central field problem to be used later on atomic systems.

A particle constrained to move in the x, y plane under the action of a restoring force directed toward the origin and proportional to the distance of the particle from the origin constitutes a two dimensional, isotropic, harmonic oscillator. The configuration space for the particle is of dimension 2, and its classical potential energy is $V = \frac{1}{2}k(x^2 + y^2)$. Conjugate to x and y are momenta p_x and p_y, and the classical kinetic energy for a particle of mass μ is $K = (p_x^2 + p_y^2)/2\mu$. Replacing p_x by $(\hbar/i)(\partial/\partial x)$ and p_y by $(\hbar/i)(\partial/\partial y)$ in this expression we have for the Hamiltonian operator

$$H = K + V = -\frac{\hbar^2}{2\mu}\left[\frac{\partial^2}{\partial x^2} + \frac{\partial^2}{\partial y^2}\right] + \tfrac{1}{2}k(x^2 + y^2)$$

The Schrödinger equation is thus $H\psi = E\psi$:

$$-\frac{\hbar^2}{2\mu}\left[\frac{\partial^2\psi}{\partial x^2} + \frac{\partial^2\psi}{\partial y^2}\right] + \tfrac{1}{2}k(x^2 + y^2)\psi = E\psi \tag{4.2}$$

Note that if we suppress y in this equation, it becomes identical with that for the one dimensional oscillator.

Equation 4.2 is solvable as it stands, but in order to get the flavor of the use of coordinate systems other than Cartesian, I shall transform it to plane polar coordinates. This can be done in a perfectly logical, mathematically elegant, distressingly complicated way that I take pleasure in avoiding. In practice all that needs to be done is to recognize in the operator

$$\frac{\partial^2}{\partial x^2} + \frac{\partial^2}{\partial y^2} \equiv \nabla^2$$

the two dimensional form in Cartesian coordinates of the Laplacian operator that occurs frequently in other branches of mathematical physics. The Schrödinger equation may thus be written symbolically as

$$-\frac{\hbar^2}{2\mu}\nabla^2\psi + \tfrac{1}{2}k\,(x^2 + y^2)\,\psi = E\psi$$

To transform to plane polar coordinates, write $x = r\cos\varphi$; $y = r\sin\varphi$ $(0 \le r \le \infty; 0 \le \varphi \le 2\pi)$ so that $x^2 + y^2 = r^2$, and look up the form taken by the Laplacian operator in plane polar coordinates in a handbook

(Appendix II). The result is

$$\nabla^2 = \frac{1}{r}\frac{\partial}{\partial r} r \frac{\partial}{\partial r} + \frac{1}{r^2}\frac{\partial^2}{\partial \varphi^2}$$

so that in plane polar coordinates the Schrödinger equation is

$$-\frac{\hbar^2}{2\mu}\left[\frac{1}{r}\frac{\partial}{\partial r} r \frac{\partial \psi}{\partial r} + \frac{1}{r^2}\frac{\partial^2 \psi}{\partial \varphi^2}\right] + \tfrac{1}{2}kr^2\psi = E\psi \tag{4.3}$$

This procedure is standard practice in quantum mechanical investigations, and the kinetic energy operator always turns out to involve some form of the Laplacian operator, so that the Schrödinger equation for a single particle of mass m may always be written

$$-\frac{\hbar^2}{2m}\nabla^2\psi + V\psi = E\psi$$

without reference to the coordinate system employed.

4.10 SEPARATION OF THE SCHRÖDINGER EQUATION

We shall solve equation 4.3 by the only known general method that leads to analytic solutions of such problems, for in general a partial differential equation cannot be solved exactly unless it can be separated. We assume a solution in the form of a product $\psi = R(r)\Phi(\varphi)$, where R is a function of r only and Φ a function of φ only. Making the indicated substitution into (4.3), we have

$$-\frac{\hbar^2}{2\mu}\left[\Phi \frac{1}{r}\frac{d}{dr} r \frac{dR}{dr} + \frac{1}{r^2} R \frac{d^2\Phi}{d\varphi^2}\right] + \tfrac{1}{2}kr^2 R\Phi = ER\Phi$$

Now divide through by $R\Phi$ and multiply by r^2. After a rearrangement,

$$\left[\frac{r}{R}\frac{d}{dr} r \frac{dR}{dr} - \frac{\mu k}{\hbar^2} r^4 + \frac{2\mu}{\hbar^2} Er^2\right] + \frac{1}{\Phi}\frac{d^2\Phi}{d\varphi^2} = 0$$

Of the terms in this equation, those enclosed by the brackets depend only on r while the last one depends only on φ. If thus we change one of the variables r, φ but hold the other fixed, it would seem possible to alter one of these groups of terms at will while the other is unaffected. But our differential equation must be satisfied for *all* values of the independent variables. It must therefore follow that each of these two groups of terms involving a single variable is separately constant independent of the value of the variable, and I shall anticipate a result shortly to be derived if I write the constant as

$-m^2$ so that

$$\frac{d^2\Phi}{d\varphi^2} = -m^2\Phi \tag{4.4}$$

and

$$r\frac{d}{dr}r\frac{dR}{dr} - \frac{\mu k}{\hbar^2}r^4R + \frac{2\mu}{\hbar^2}Er^2R - m^2R = 0 \tag{4.5}$$

Now according to Section 3.5, the angular equation 4.4 has solutions $\Phi = \exp(im\varphi)$. To make the wave function single valued and continuous, m must be purely real and an integer $m = 0, \pm 1, \pm 2, \ldots$, since otherwise Φ will not repeat itself as φ increases through 2π. A similar study of the radial equation 4.5 shows properties analogous to those displayed by equation 3.10 for the linear, harmonic oscillator: solutions R that are finite, continuous, and normalizable exist only if E is restricted to the discrete set of values

$$E = (n + |m| + 1)h\nu$$

in which n is a positive, even integer $n = 0, 2, 4, \ldots$. When this is done, the radial functions turn out to be of the form

$$R = \left(\text{polynomial in } \frac{r}{x_0}\right) \exp\left(-\frac{1}{2}\frac{r^2}{x_0^2}\right)$$

with $x_0 = (\hbar/\sqrt{\mu k})^{1/2}$. The discrete sets of numbers n, m are known as *quantum numbers*.

This technique of assuming a wave function in the form of a product of functions of a single variable and thereby breaking up the Schrödinger equation into several ordinary differential equations is called "separating" a partial differential equation. The choice of a coordinate system plays an important role in separating an equation, for a Schrödinger equation separable in one coordinate system may not be separable in another; and it is part of the craft of the investigator to recognize coordinate systems appropriate for a given problem.

4.11 SCALAR PRODUCTS AND NORMALIZATION

By judicious use of a handbook, we have been able to make a speedy transition from Cartesian to plane polar coordinates in the treatment of our Schrödinger equation. The same handbook will also enable us to straighten out certain details in the calculation of normalization and other integrals in this coordinate system.

The general notation $\int \phi^*\psi \, d\tau$ for the scalar product of two "vectors" in function space is still valid, for it implies integration of the product $\phi^*\psi$ over the entirety of the configuration space of the system. When the axes of configuration space are Cartesian, the limits of integration are $-\infty$ to $+\infty$ for each Cartesian axis, and the volume element $d\tau$ has the interpretation

$$d\tau = dx_1\, dy_1\, dz_1 \cdots dx_N\, dy_N\, dz_N$$

We have already had in Section 3.7 an example in the one dimensional harmonic oscillator of the special case $d\tau = dx_1$. The limits of integration for coordinates other than Cartesian are, however, different, and a good handbook should indicate the range of numerical values over which a coordinate can vary. (Unfortunately, handbooks do not always do so.) For plane polar coordinates, the angle φ can range from 0 to 2π and the radius r from 0 to ∞, and these must be the limits of integration over configuration space.

Another feature of non-Cartesian coordinate systems is that the volume element $d\tau$ always includes functions of the coordinates themselves as well as their differentials. For plane polar coordinates, the volume element is

$$d\tau = r\, dr\, d\varphi$$

and we shall shortly have examples of similar modifications of the volume element in other coordinate systems. These changes in $d\tau$ are purely geometric in character and have nothing to do with quantum mechanics. For the reason for them, the reader is referred to texts on tensor analysis.[11,12] In practice all he needs to do is look up the needed volume element in a handbook.

As examples of the use of these rules, I shall apply them to the normalization of the wave function for the two dimensional, isotropic, harmonic oscillator. We have already factored the wave function $\psi = R(r)\Phi(\varphi)$, and our task is to multiply this product by a suitable normalization constant C such that

$$\int \psi^*\psi\, d\tau = \int_0^\infty r\, dr \int_0^{2\pi} d\varphi C^*CR^*R\Phi^*\Phi = 1$$

R^*R and $\Phi^*\Phi$ will always be real, so that C may also be chosen to be real. Rearranging the integral we have

$$C^2 \int_0^\infty R^*Rr\, dr \int_0^{2\pi} \Phi^*\Phi\, d\varphi = 1$$

whence

$$C = \left\{ \int_0^\infty R^*Rr\, dr \int_0^{2\pi} \Phi^*\Phi\, d\varphi \right\}^{-1/2}$$

The integration over the angle is easily performed:

$$\int_0^{2\pi} \Phi^*\Phi\, d\varphi = \int_0^{2\pi} e^{-im\varphi} e^{im\varphi}\, d\varphi = \int_0^{2\pi} d\varphi = 2\pi$$

To integrate over r, the functions satisfying the radial equation must be investigated. They will contain the quantum numbers n and m as parameters, and I shall write them $R_{nm}(r)$. Examples of the R_{nm} are

$$R_{oo} = \exp\left[-\frac{1}{2}\left(\frac{r}{x_0}\right)^2\right]$$

$$R_{om} = \left(\frac{r}{x_0}\right)^{|m|}\exp\left[-\frac{1}{2}\left(\frac{r}{x_0}\right)^2\right]$$

$$R_{20} = \left[1 - \left(\frac{r}{x_0}\right)^2\right]\exp\left[-\frac{1}{2}\left(\frac{r}{x_0}\right)^2\right]$$

$$R_{2\pm 1} = \left(\frac{r}{x_0}\right)\left[1 - \frac{1}{2}\left(\frac{r}{x_0}\right)^2\right]\exp\left[-\frac{1}{2}\left(\frac{r}{x_0}\right)^2\right]$$

Because they are real, $R^*R = R^2$, so that the normalization integral for R_{oo} is

$$\int_0^\infty R_{oo}{}^2 r\, dr = \int_0^\infty r e^{-r^2/x_0{}^2}\, dr = \tfrac{1}{2}x_0{}^2$$

Complete normalization of the ground state wave function then requires

$$C = (2\pi)^{-1/2}(\tfrac{1}{2}x_0{}^2)^{-1/2} = (\pi x_0{}^2)^{-1/2}$$

whence, finally, the normalized wave function for the ground state of the two dimensional, isotropic, harmonic oscillator in plane polar coordinates is

$$\psi_{00} = (\pi x_0{}^2)^{-1/2}\exp\left[-\frac{1}{2}\left(\frac{r}{x_0}\right)^2\right]$$

Associated with this wave function is an energy $E_0 = h\nu$.

PROBLEM

4.1 (1) For the two dimensional, isotropic, harmonic oscillator normalize the wave functions for the second excited state

$$\psi_{20} \sim \left[1 - \left(\frac{r}{x_0}\right)^2\right]\exp\left[-\frac{1}{2}\left(\frac{r}{x_0}\right)^2\right]$$

$$\psi_{02} \sim \left(\frac{r}{x_0}\right)^2\exp\left[-\frac{1}{2}\left(\frac{r}{x_0}\right)^2\right]e^{2i\varphi}$$

$$\psi_{0-2} \sim \left(\frac{r}{x_0}\right)^2\exp\left[-\frac{1}{2}\left(\frac{r}{x_0}\right)^2\right]e^{-2i\varphi}$$

(2) Show that ψ_{02} and ψ_{0-2} are orthogonal.

4.12 DEGENERACY

Our calculations for the two dimensional, isotropic, harmonic oscillator have yielded the following information concerning the wave functions. For every distinct pair of quantum numbers n, m drawn from the list

$$\begin{aligned} n &= 0, 2, 4, \ldots \\ m &= 0, \pm 1, \pm 2, \ldots \end{aligned} \tag{4.6}$$

there is a distinct wave function ψ_{nm}. Associated with each wave function is an energy level

$$E = (n + |m| + 1)h\nu$$

It is evident that with the exception of the ground state $n = m = 0$, all of the energy levels are degenerate, for there are many ways in which pairs n, m can be drawn from the list (4.6) which leave $n + |m|$ unaltered. Thus $n = 0$; $m = 1, -1$ belong to the doubly degenerate level $E = 2h\nu$; and to the level $E = 3h\nu$ are assigned the three possibilities

$$\begin{aligned} n &= 2; \qquad m = 0 \\ n &= 0; \qquad m = 2 \\ n &= 0; \qquad m = -2 \end{aligned}$$

By extending this classification the student will speedily convince himself that the energy level $E = (n + |m| + 1)h\nu$ is $n + |m| + 1$ fold degenerate.

Vector algebra extended to function spaces then tells us that wave functions associated with different energy levels will automatically be orthogonal, but that wave functions associated with the same energy level will be orthogonal only if we choose to make them so. Furthermore in the absence of other information, any orthogonal, degenerate subset is equivalent to any other.

As an example, consider the doubly degenerate pair of functions

$$\begin{aligned} \psi_{01} &= (\pi x_0^2)^{-1/2}\left(\frac{r}{x_0}\right) \exp\left[-\frac{1}{2}\left(\frac{r}{x_0}\right)^2\right] e^{i\varphi} \\ \psi_{0-1} &= (\pi x_0^2)^{-1/2}\left(\frac{r}{x_0}\right) \exp\left[-\frac{1}{2}\left(\frac{r}{x_0}\right)^2\right] e^{-i\varphi} \end{aligned} \tag{4.7}$$

They happen to be already orthogonal, for the product $\psi_{0-1}^* \psi_{01}$ contains a factor $(e^{-i\varphi})^* e^{i\varphi} = e^{2i\varphi}$, and integration of $e^{2i\varphi}$ from 0 to 2π yields 0. Any normalized linear combination of ψ_{01} and ψ_{0-1}, however, is also a suitable wave function; in particular, the mutually orthogonal linear combinations

$$\psi_x = \frac{1}{\sqrt{2}}\,\psi_{01} + \frac{1}{\sqrt{2}}\,\psi_{0-1} = \sqrt{2}\,(\pi x_0^{\,2})^{-1/2}\left(\frac{r}{x_0}\right)\cos\varphi\,\exp\left[-\frac{1}{2}\left(\frac{r}{x_0}\right)^2\right]$$

$$\psi_y = -\frac{i}{\sqrt{2}}\,\psi_{01} + \frac{i}{\sqrt{2}}\,\psi_{0-1} = \sqrt{2}\,(\pi x_0^{\,2})^{-1/2}\left(\frac{r}{x_0}\right)\sin\varphi\,\exp\left[-\frac{1}{2}\left(\frac{r}{x_0}\right)^2\right] \tag{4.8}$$

have the pictorial advantage that the resulting wave functions are real so that we may draw contour plots of them (Figure 4.4).

The diagrams of Figure 4.4 show schematically contour lines of constant ψ for the ground and first excited states. The ground state ψ_{00} is radially symmetric and the wave function is everywhere positive. ψ_x has a maximum and a minimum located on the x axis, and there is a nodal line coincident with the y axis. ψ_y is identical with ψ_x but rotated through an angle of 90°, so that its nodal line is coincident with the x axis.

Before letting his eye glide too swiftly over the coefficients used in the linear combinations on the left of equations 4.8, the student should note that these four numbers constitute the components of the unitary matrix M which was introduced in equation 1.10. Orthogonal or unitary matrices always occur when making a transformation from one orthogonal, degenerate set of wave functions to another.

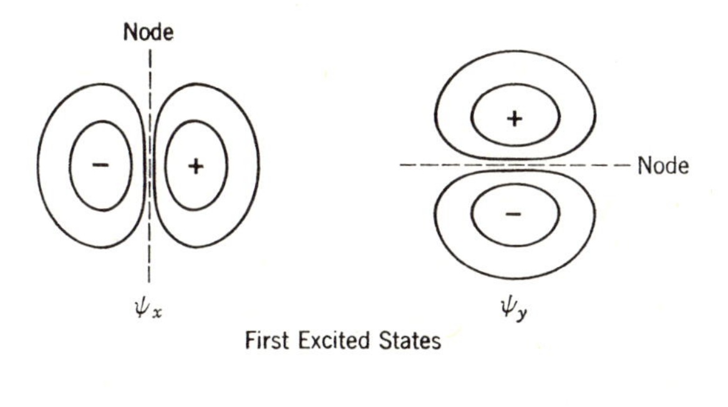

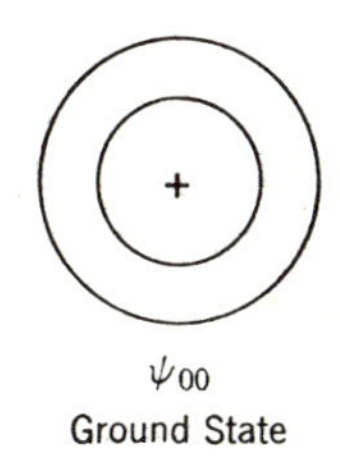

Figure 4.4 Wave functions for the two dimensional, isotropic, harmonic oscillator. The contours join together points for which ψ has the same numerical value after the manner of geologic altitude maps.

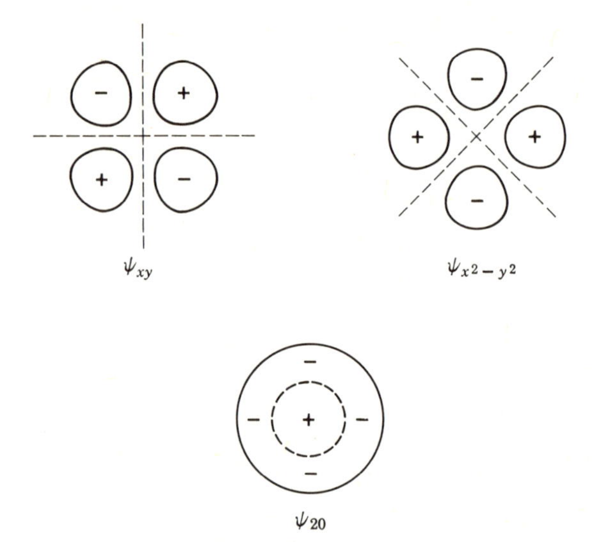

Figure 4.5 Degenerate second excited states of the two dimensional, isotropic, harmonic oscillator. Nodes are indicated by dashed lines.

The orientation of the real wave functions along the x and y axes is furthermore not unique, for we may take additional orthogonal linear combinations

$$\begin{aligned} \psi_1 &= (\cos\chi)\psi_x + (\sin\chi)\psi_y \\ \psi_2 &= (-\sin\chi)\psi_x + (\cos\chi)\psi_y \end{aligned} \tag{4.9}$$

for χ arbitrary and still retain a valid pair of solutions for the first excited state of the oscillator. The coefficients appearing in (4.9) are the components of the general, orthogonal, rotation matrix used in Problem 1.11 and have the effect of rotating the patterns of Figure 4.4 through an angle χ. This arbitrariness in the positions of the nodes of a degenerate pair of wave functions is reminiscent of the similar arbitrariness in the nodes of the degenerate Hückel molecular orbitals $\boldsymbol{\psi}_2$ and $\boldsymbol{\psi}_3$ of benzene which were discussed in Section 2.6.

Continuing this line of investigation to the second excited state for which $E = 3h\nu$, we find that there are three degenerate wave functions ψ_{20}, ψ_{02}, ψ_{0-2}, of which the last two are complex. When appropriate orthogonal linear combinations of them are taken, three real wave functions result whose contours have the symmetry properties shown in Figure 4.5.

The student should recall at this point that the probability distribution of the particle in its configuration space is $\psi^*\psi$ (ψ^2 if ψ is real) and ponder the significance of these diagrams in terms of the probability distribution of the particle.

PROBLEM

4.2 Construct the set of linear combinations of ψ_{20}, ψ_{02}, ψ_{0-2} which are real and orthogonal and whose contours are those of Figure 4.5. As a starting point, see Problem 4.1. Remember that the coefficients used in your linear combinations must be the components of a unitary matrix.

4.13 ORTHOGONALITY AND SYMMETRY

There is a suggestive relationship between the nodal contours of a wave function and its orthogonality to other wave functions, and it will repay the student with the aid of Figure 4.4 to picture geometrically the result of multiplying ψ_x times ψ_{00} and integrating over the configuration space. Because ψ_{00} is circularly symmetric and positive everywhere, the node contained in ψ_x will persist into the product $\psi_x\psi_{00}$, which will then be negative on the left side of the y axis and positive on the right. Furthermore, the result of integrating over the left half space will be identical in magnitude but opposite in sign from the contribution to the integral from the right half space, so that the sum of the two contributions vanishes:

$$\int \psi_x \psi_{00} \, d\tau = 0$$

Of course, the fact that ψ_x and ψ_{00} belong to two different energy levels assures us that they should be orthogonal even without this geometric argument, but in practical calculations it frequently repays the analyst to organize wave functions into symmetry classes in such a way that members of different classes are automatically orthogonal just from the way their nodal contours are arranged.

As an exercise, the student should satisfy himself from a study of the contour diagrams of Figure 4.4 for the degenerate pair ψ_x and ψ_y that because of their different symmetries

$$\int \psi_x \psi_y \, d\tau = 0$$

4.14 ANGULAR MOMENTUM

An important theorem in classical mechanical problems that deal with rotational motion or with the motion of a system of particles in a central field requires the conservation of angular momentum of the system. This theorem has as far reaching an effect on the motion as does the law of conservation of energy, and in its original form it was first noticed empirically by Kepler in his researches on planetary motion, where he stated it as his law of "equal areas—equal times." The quantum mechanical analog to the classical theorem lies in the fact that the angular momentum operator which we shall shortly set up commutes with the Hamiltonian. From what we have learned of the significance of this fact in vector algebra, both the angular momentum and Hamiltonian operators must then have eigenfunctions in common, meaning experimentally that both energy and angular momentum may simultaneously have definite values.

The classical mechanical picture of the angular momentum is of a vector quantity lying along the axis of rotation of a particle. (This distinguishes it from the energy, which is a scalar.) We could take a good deal of time at this point defining what is meant by an axis of rotation, but for our present application to the two dimensional, harmonic oscillator, the axis of rotation is drawn from the origin perpendicular to the xy plane in which the motion takes place. It therefore has only the single component $\mathscr{L}_z$ along the z axis, with no components in the xy plane. In Cartesian coordinates, its classical magnitude is

$$\mathscr{L}_z = xp_y - yp_x$$

showing that it is a function of both position coordinates and their conjugate momenta.

In plane polar coordinates, $\mathscr{L}_z$ is the momentum conjugate to φ, and after transformation from Cartesian coordinates,* its corresponding (Hermitian)

* The transformation is carried out as follows. Into the classical Cartesian formula substitute $p_x = (\hbar/i)(\partial/\partial x)$; $p_y = (\hbar/i)(\partial/\partial y)$, so that

$$\mathscr{L}_z = \frac{\hbar}{i}\left(x\frac{\partial}{\partial y} - y\frac{\partial}{\partial x}\right)$$

But

$$\frac{\partial}{\partial x} = \left(\frac{\partial r}{\partial x}\right)_y \frac{\partial}{\partial r} + \left(\frac{\partial \varphi}{\partial x}\right)_y \frac{\partial}{\partial \varphi}$$

and

$$\frac{\partial}{\partial y} = \left(\frac{\partial r}{\partial y}\right)_x \frac{\partial}{\partial r} + \left(\frac{\partial \varphi}{\partial y}\right)_x \frac{\partial}{\partial \varphi}$$

With

$$r = \sqrt{x^2 + y^2}$$

and

$$\varphi = \arctan(y/x)$$

operator is $\mathscr{L}_z = (\hbar/i)(\partial/\partial\varphi)$. When we examine the Hamiltonian operator H in plane polar coordinates equation 4.3, we obtain the commutation rule

$$\mathscr{L}_z H = H\mathscr{L}_z$$

or

$$\left\{\frac{\hbar}{i}\frac{\partial}{\partial\varphi}\right\}\left\{-\frac{\hbar^2}{2\mu}\left[\frac{1}{r}\frac{\partial}{\partial r}r\frac{\partial}{\partial r}+\frac{1}{r^2}\frac{\partial^2}{\partial\varphi^2}\right]+\tfrac{1}{2}kr^2\right\}\psi$$
$$=\left\{-\frac{\hbar^2}{2\mu}\left[\frac{1}{r}\frac{\partial}{\partial r}r\frac{\partial}{\partial r}+\frac{1}{r^2}\frac{\partial^2}{\partial\varphi^2}\right]+\tfrac{1}{2}kr^2\right\}\left\{\frac{\hbar}{i}\frac{\partial}{\partial\varphi}\right\}\psi$$

which is valid for any ψ whether ψ is a wave function or not. Our theorem then states that there exists a set of eigenfunctions of H which are simultaneously eigenfunctions of $\mathscr{L}_z$. Upon examination of our wave functions in the form $\psi = R_{nm}(r)\exp(im\varphi)$, it readily appears that

$$\mathscr{L}_z\psi = \frac{\hbar}{i}\frac{\partial}{\partial\varphi}R_{nm}(r)e^{im\varphi} = m\hbar R_{nm}(r)e^{im\varphi} = m\hbar\psi \tag{4.10}$$

Thus when we use a complex exponential form for the angular part of the wave function, ψ is an eigenfunction of $\mathscr{L}_z$ with eigenvalue $m\hbar$. This is *not* the case when the real wave functions (4.8) involving $\sin m\varphi$ and $\cos m\varphi$ are used, for these linear combinations of complex exponentials are not eigenfunctions of $\mathscr{L}_z = (\hbar/i)(\partial/\partial\varphi)$ and the student should convince himself of this fact by a direct calculation.

According to the discussion of postulate 4 (Section 4.6) when the two dimensional, isotropic, harmonic oscillator is in a state ψ whose angular factor is a complex exponential $e^{im\varphi}$, both the energy and the angular momentum have definite values simultaneously:

$$\langle\mathscr{L}_z\rangle = \int \psi^* \frac{\hbar}{i}\frac{\partial\psi}{\partial\varphi}\,d\tau = m\hbar$$

$$E = \int \psi^* H\psi\,d\tau = (n + |m| + 1)h\nu$$

---

we have

$$\left(\frac{\partial r}{\partial x}\right)_y = \frac{x}{\sqrt{x^2+y^2}} = \frac{x}{r} = \cos\varphi; \qquad \left(\frac{\partial r}{\partial y}\right)_x = \sin\varphi$$

$$\left(\frac{\partial\varphi}{\partial x}\right)_y = -\frac{y}{x^2+y^2} = -\frac{1}{r}\sin\varphi; \qquad \left(\frac{\partial\varphi}{\partial y}\right)_x = \frac{1}{r}\cos\varphi$$

After substitution into $\mathscr{L}_z$ and simplification there remains

$$\mathscr{L}_z = \frac{\hbar}{i}\frac{\partial}{\partial\varphi}$$

but if it is in a state ψ whose angular factor is either sin $m\varphi$ or cos $m\varphi$, only the energy has a definite value. The student may well ask, "How does the oscillator know whether or not it should be in a state ψ for which the angular momentum is definite?" The answer lies in the type of experiment chosen to observe the system, for the experiment will disturb the system in such a way as to call forth the appropriate wave function. When we come to apply these ideas to atoms rather than to oscillators, we shall see that if we disturb an atom by an experiment which places it in a uniform magnetic field (Zeeman effect), the experiment will force the atom into a state for which its angular momentum is definite, but that if we disturb the atom by bringing up another atom or group of atoms with which it can combine chemically, the angular momentum is usually indefinite. Thus the very act of performing an experiment determines the type of information which will be obtained. The only remaining question is philosophic: what is the quantum mechanical system doing when we are not observing it?, but the modern physicist is content to leave this metaphysical question to the philosophers, the theologians, and (with reluctance) to an occasional politician.

PROBLEMS

4.3 In Cartesian coordinates construct the Hamiltonian operator for the two dimensional, isotropic, harmonic oscillator and show that it commutes with the angular momentum operator

$$\mathscr{L}_z = xp_y - yp_x$$

4.4 The Schrödinger equation (4.2) for the two dimensional, isotropic, harmonic oscillator is separable in Cartesian coordinates. Separate the equation and solve the separated parts (*hint:* see Section 3.7). Write down a general expression for the energy levels and determine the degeneracy of each level.

4.5 In Cartesian coordinates the doubly degenerate first excited state of the two dimensional, isotropic oscillator has an energy $E = 2h\nu$ and a pair of wave functions

$$\psi_x = (\pi x_0^2)^{-1/2}\sqrt{2}\left(\frac{x}{x_0}\right)\exp\left[-\frac{1}{2}\frac{x^2+y^2}{x_0^2}\right]$$

$$\psi_y = (\pi x_0^2)^{-1/2}\sqrt{2}\left(\frac{y}{x_0}\right)\exp\left[-\frac{1}{2}\frac{x^2+y^2}{x_0^2}\right]$$

(1) Show that these wave functions are *not* eigenfunctions of the angular momentum operator $\mathscr{L}_z$, where $\mathscr{L}_z$ is expressed in Cartesian coordinates.

(2) Construct orthonormal linear combinations of ψ_x and ψ_y which *are* eigenfunctions of $\mathscr{L}_z$ and find their eigenvalues with respect to $\mathscr{L}_z$.

4.6 Calculate the expectation $\langle \mathscr{L}_z \rangle$ of the angular momentum for the states ψ_x and ψ_y (equations 4.8). How do you account for this result in view of the fact that the only possible outcomes of a single experiment measuring $\mathscr{L}_z$ must for the first excited state of the oscillator be $\pm\hbar$?

4.7 The electron-in-a-ring problem can be visualized as the motion of an electron in a loop of wire bent into a circle. When treated in plane polar coordinates, the radial position of the electron is fixed at $r =$ the radius of the ring, so that the wave function varies only with φ. For convenience we set the potential energy of the electron in the ring at $V = 0$. (On a molecular scale this is a crude model of benzene.)

(1) Set up the Schrödinger equation for the electron-in-a-ring in plane polar coordinates and integrate it to find the energy levels and normalized wave functions.

(2) Prove that the angular momentum operator $\mathscr{L}_z$ commutes with the Hamiltonian operator for the ring and find the eigenvalues of $\mathscr{L}_z$.

REFERENCES

1. J. L. Synge and B. A. Griffith, *Principles of Mechanics*, McGraw-Hill, Inc., New York, 1959, pp. 430–435.
2. E. T. Whittaker, *A Treatise on Analytical Dynamics*, Dover Press, New York, 1944, p. 54, Chapter X.
3. L. Pauling and E. B. Wilson, Jr., *Introduction to Quantum Mechanics*, McGraw-Hill, Inc., New York, 1935, p. 76.
4. W. Kauzmann, *Quantum Chemistry*, Academic Press, Inc., New York, 1957, p. 206.
5. H. Eyring, J. Walter, and G. E. Kimball, *Quantum Chemistry*, John Wiley and Sons, Inc., New York, 1944, p. 27.
6. W. Kauzmann, *loc. cit.*, p. 112.
7. H. Margenau and G. M. Murphy, *The Mathematics of Physics and Chemistry*, D. Van Nostrand, Inc., New York, 1956, Vol. I, p. 344.
8. P. M. Morse and H. Feshbach, *Methods of Theoretical Physics*, McGraw-Hill, Inc., New York, 1953, pp. 771–778.
9. P. M. Morse and H. Feshbach, *ibid.*, Chapter 5.
10. H. Margenau and G. M. Murphy, *loc. cit.*, Chapter 2.
11. H. Margenau and G. M. Murphy, *loc. cit.*, Chapter 5.
12. H. Lass, *Vector and Tensor Analysis*, McGraw-Hill, Inc., New York, 1950.

Chapter 5

THE CENTRAL FIELD PROBLEM

5.1 DEFINITION

The central field problem in both classical and quantum mechanics is the study of the motion of a single particle constrained to move in a field of force which is a function only of the distance $r = (x^2 + y^2 + z^2)^{1/2}$ from the origin. In two dimensional form, the problem of the isotropic oscillator considered in Chapter 4 is of this type, and we shall find that much of what we learn here falls into line with our experience there.

The Schrödinger equations of all central field problems are separable in spherical polar coordinates. Furthermore, the differential equations involving the angular coordinates are always the same, so that the factored wave functions $\psi = R(r)Y(\theta, \varphi)$ always have the same angular part $Y(\theta, \varphi)$ and differ only in the radial portion $R(r)$. All calculations involving the angles only are therefore the same for all central field problems.

In practice, the most important central field problem is that of the hydrogen atom, but to the extent that a polyelectronic atom can be considered as a central field problem (and this is only approximately true), all of what we develop in this chapter goes over to that case also.

5.2 SPHERICAL POLAR COORDINATES

A diagram of the spherical polar coordinate system is sketched in Figure 5.1. A point P is located in this system by a triple of numbers r, θ, φ in which r is the distance from the origin, θ is the "polar angle" running from 0 to π which corresponds on the surface of a globe to geographic latitude, and φ running from 0 to 2π is the "azimuthal angle" corresponding to geographic longitude. The relation to Cartesian axes x, y, z is

$$\begin{aligned} x &= r \sin\theta \cos\varphi \\ y &= r \sin\theta \sin\varphi \\ z &= r \cos\theta \end{aligned}$$

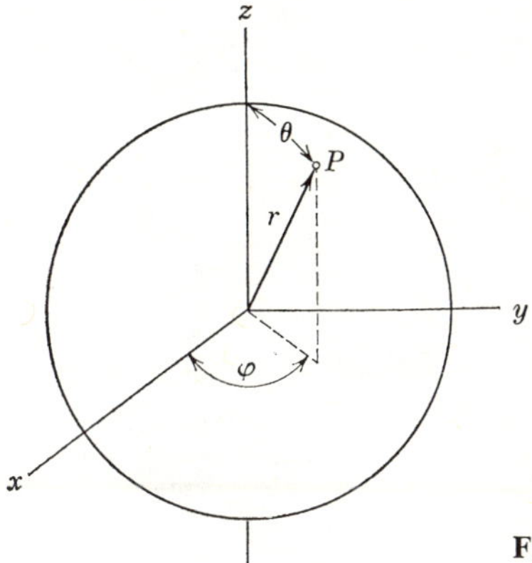

Figure 5.1 The spherical polar coordinate system.

and the volume element $d\tau$ which in Cartesian coordinates is

$$d\tau = dx\, dy\, dz$$

in spherical polar coordinates becomes

$$d\tau = r^2 \sin\theta\, dr\, d\theta\, d\varphi$$

An integral over configuration space for a single particle thus has the form

$$\int \Phi^* \psi\, d\tau = \int_0^\infty r^2\, dr \int_0^\pi \sin\theta\, d\theta \int_0^{2\pi} d\varphi\, [\Phi^* \psi]$$

which is the definition of the "scalar product" between two "vectors" Φ and ψ. Finally, the Laplacian operator ∇^2 which in Cartesian coordinates is

$$\nabla^2 = \frac{\partial^2}{\partial x^2} + \frac{\partial^2}{\partial y^2} + \frac{\partial^2}{\partial z^2}$$

now becomes

$$\nabla^2 = \frac{1}{r^2}\frac{\partial}{\partial r} r^2 \frac{\partial}{\partial r} + \frac{1}{r^2 \sin\theta}\frac{\partial}{\partial\theta} \sin\theta \frac{\partial}{\partial\theta} + \frac{1}{r^2 \sin^2\theta}\frac{\partial^2}{\partial\varphi^2} \tag{5.1}$$

5.3 THE SCHRÖDINGER EQUATION

To construct the Hamiltonian operator for a particle of mass μ we need the classical kinetic energy in Cartesian coordinates

$$K = \frac{p_x^2 + p_y^2 + p_z^2}{2\mu}$$

together with the classical potential energy $V[(x^2 + y^2 + z^2)^{1/2}] = V(r)$. Replacing p_x by $(\hbar/i)(\partial/\partial x)$, and so on, the kinetic energy operator proves to be

$$K = -\frac{\hbar^2}{2\mu}\left[\frac{\partial^2}{\partial x^2} + \frac{\partial^2}{\partial y^2} + \frac{\partial^2}{\partial z^2}\right]$$
$$= -\frac{\hbar^2}{2\mu}\nabla^2$$

Converting to spherical polar coordinates by use of (5.1), we set up the Schrödinger equation $H\psi = (K + V)\psi = E\psi$:

$$-\frac{\hbar^2}{2\mu}\left\{\frac{1}{r^2}\frac{\partial}{\partial r}r^2\frac{\partial\psi}{\partial r} + \frac{1}{r^2\sin\theta}\frac{\partial}{\partial\theta}\sin\theta\frac{\partial\psi}{\partial\theta} + \frac{1}{r^2\sin^2\theta}\frac{\partial^2\psi}{\partial\varphi^2}\right\} + V(r)\psi = E\psi \qquad (5.2)$$

At this point I begin to skip over some details, relying heavily on the analogous operations of Sections 4.10 to 4.11. When we seek a solution of equation 5.2 in the form $\psi = R(r)P(\theta)\Phi(\varphi)$, the equation separates into three ordinary differential equations in each of the three variables, r, θ, φ:

$$\frac{1}{r^2}\frac{d}{dr}r^2\frac{dR}{dr} + \frac{2\mu}{\hbar^2}[E - V(r)]R - \frac{\lambda}{r^2}R = 0 \qquad (5.3)$$

$$\frac{1}{\sin\theta}\frac{d}{d\theta}\sin\theta\frac{dP}{d\theta} + \lambda P - \frac{m^2}{\sin^2\theta}P = 0 \qquad (5.4)$$

$$\frac{d^2\Phi}{d\varphi^2} = -m^2\Phi \qquad (5.5)$$

There are several things to notice about these equations. The first is the presence of the separation constants λ and m. These arise in the same way as did the separation constant m in the treatment of the two dimensional, isotropic oscillator—the several terms in the original partial differential equation may be grouped into units each of which is a function of a single variable only, and each unit must therefore be separately equal to a constant if the equation is to be satisfied for all values of r, θ, φ. Second, note that the potential energy $V(r)$ occurs only in the radial equation 5.3, so that the angular equations 5.4 and 5.5 are the same for all central field problems. Finally, the energy eigenvalue E which must be obtained by solution of the radial equation 5.3 involves λ as a parameter but not m, so that the energy levels will be independent of m. This immediately suggests the presence of degeneracy in all central field problems.

The solution of equation 5.5 is from equation 4.4 by now familiar:

$$\Phi = \exp(im\varphi)$$

where in order for ψ to be single valued and continuous, m is restricted to be an integer, positive, negative, or 0. This is our first quantum number.

The solutions of the θ equation 5.4 which satisfy the quantum restrictions of continuity, finiteness, and single valuedness are known in mathematics as the associated Legendre polynomials $P_l^{|m|}(\cos\theta)$, although they are actually not polynomials in $\cos\theta$, for they possess a factor $\sin^{|m|}\theta$. Lists of them will be found in standard texts.[1,2]

The second quantum number l must be an integer drawn from the list $l = 0, 1, 2, \ldots$; then in terms of l it turns out that $\lambda = l(l+1) = 0, 2, 6, \ldots$. An important feature of the solutions to equation 5.4 is that a restriction is imposed on the upper value of $|m| \le l$, so that in summary the quantum numbers m and l may be chosen from the sets

$$l = 0, 1, 2, \ldots$$

$$m = -l, -l+1, \ldots, 0, \ldots, l-1, l$$

Because E is independent of m, the student should confirm from this list that for each value of l there will be an energy level that is $2l+1$ fold degenerate.

I ignore the solution of the radial equation 5.3 for the time being and summarize what we have learned of the wave function by writing

$$\psi = R(r)Y_{lm}(\theta, \varphi)$$

for which $R(r)$ is as yet unknown. When normalized over the angular coordinates in the sense

$$\int_0^{\pi} \sin\theta\, d\theta \int_0^{2\pi} d\varphi [Y_{lm}^* Y_{lm}] = 1$$

explicit expressions for the first Y_{lm} are

$$l = 0; \quad m = 0; \quad Y_{00} = \frac{1}{\sqrt{2\pi}}\frac{1}{\sqrt{2}}$$

$$l = 1; \quad m = -1; \quad Y_{1-1} = \frac{1}{\sqrt{2\pi}}\sqrt{\tfrac{3}{4}}\sin\theta\exp(-i\varphi)$$

$$m = 0; \quad Y_{10} = \frac{1}{\sqrt{2\pi}}\sqrt{\tfrac{3}{2}}\cos\theta$$

$$m = 1; \quad Y_{11} = \frac{1}{\sqrt{2\pi}}\sqrt{\tfrac{3}{4}}\sin\theta\exp(i\varphi) \tag{5.6}$$

$$l = 2; \quad m = -2; \quad Y_{2-2} = \frac{1}{\sqrt{2\pi}} \sqrt{\tfrac{15}{16}} \sin^2 \theta \exp(-2i\varphi)$$

$$m = -1; \quad Y_{2-1} = \frac{1}{\sqrt{2\pi}} \sqrt{\tfrac{15}{4}} \sin \theta \cos \theta \exp(-i\varphi)$$

$$m = 0; \quad Y_{20} = \frac{1}{\sqrt{2\pi}} \sqrt{\tfrac{5}{8}} (3 \cos^2 \theta - 1)$$

$$m = 1; \quad Y_{21} = \frac{1}{\sqrt{2\pi}} \sqrt{\tfrac{15}{4}} \sin \theta \cos \theta \exp(i\varphi)$$

$$m = 2; \quad Y_{22} = \frac{1}{\sqrt{2\pi}} \sqrt{\tfrac{15}{16}} \sin^2 \theta \exp(2i\varphi)$$

PROBLEM

5.1 Separate equation 5.2 to obtain equations 5.3 to 5.5.

5.4 DEGENERACY OF THE ANGULAR FUNCTIONS

The degeneracy with respect to m of the energy levels of the central field problem permits an arbitrariness in the way in which the wave functions for each degenerate level are to be chosen. We ran into this phenomenon in Section 4.12 where from the complex wave functions (4.7) for the isotropic oscillator we synthesized linear combinations (4.8) which were real. In an exactly analogous way, linear combinations of the Y_{lm} (5.6) may be formed for a fixed value of l in such a way that the resulting set of angular functions is both orthogonal and real. Thus the linear combinations

$$\begin{aligned} Y_x &= \frac{1}{\sqrt{2}} Y_{11} + \frac{1}{\sqrt{2}} Y_{1-1} = \frac{1}{\sqrt{2\pi}} \sqrt{\tfrac{3}{2}} \sin \theta \cos \varphi \\ Y_y &= -\frac{i}{\sqrt{2}} Y_{11} + \frac{i}{\sqrt{2}} Y_{1-1} = \frac{1}{\sqrt{2\pi}} \sqrt{\tfrac{3}{2}} \sin \theta \sin \varphi \\ Y_z &= Y_{10} = \frac{1}{\sqrt{2\pi}} \sqrt{\tfrac{3}{2}} \cos \theta \end{aligned} \tag{5.7}$$

are real, orthogonal, and have the property of achieving maximum and minimum values along each of the designated Cartesian axes. There are, furthermore, three mutually perpendicular nodal planes—the xy, xz, and yz planes—along each of which one of the angular factors Y_x, Y_y, Y_z (hence the complete wave function ψ) vanishes.

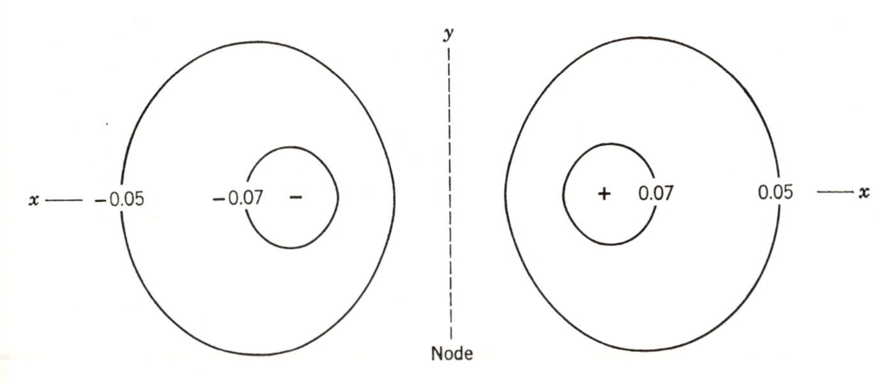

Figure 5.2 A cross section of the ψ_{2px} wave function of hydrogen.

Because we do not as yet possess explicit expressions for the radial factor $R(r)$ of the wave function, it is not possible at this point to construct a three dimensional model showing this geometry quantitatively, but if I anticipate results shortly to be derived for the hydrogen atom, a cross section of the $\psi_{2px} = R_{21}(r)\,Y_x$ wave function of hydrogen is drawn in Figure 5.2. The three dimensional contours would be those generated by revolving this figure about the x axis, and the student will perceive from this that the function must then vanish everywhere in the yz plane.

As with the transition from equations 4.7 to equations 4.8, let the student also observe that the coefficients used in setting up the linear combinations 5.7 are the components of a unitary matrix, and that the operations 5.7 that transform one orthonormal, degenerate set of functions into another are formally summarized by

$$\begin{pmatrix} Y_x \\ Y_y \\ Y_z \end{pmatrix} = \begin{pmatrix} 1/\sqrt{2} & 1/\sqrt{2} & 0 \\ -i/\sqrt{2} & i/\sqrt{2} & 0 \\ 0 & 0 & 1 \end{pmatrix} \begin{pmatrix} Y_{11} \\ Y_{1-1} \\ Y_{10} \end{pmatrix}$$

5.5 ANGULAR MOMENTA

In the course of our study of the two dimensional, isotropic oscillator, we noted in Section 4.14 that corresponding to the conservation of angular momentum theorem of classical mechanics, the quantum mechanical angular momentum operator commutes with the Hamiltonian operator. The response of general theorems in vector and matrix algebra is immediate: both operators have eigenfunctions in common, so that there exist states in which both the total energy and total angular momentum have definite values. Similar theorems exist for the central field problem, but because the motion takes place in a laboratory space of three dimensions, the details are more complicated. For motion in the xy plane only, the classical "axis of rotation" is identical with the z axis, and the angular momentum has consequently only the single component $\mathscr{L}_z$. In three dimensional space, the axis of rotation is still perpendicular to the "plane of rotation," but as the latter may have any orientation, there are three Cartesian components of the classical angular momentum $\boldsymbol{\mathscr{L}} = (\mathscr{L}_x, \mathscr{L}_y, \mathscr{L}_z)$. They are related to the Cartesian coordinates x, y, z and their conjugate momenta by

$$\mathscr{L}_x = yp_z - zp_y$$
$$\mathscr{L}_y = zp_x - xp_z$$
$$\mathscr{L}_z = xp_y - yp_x$$
$$\boldsymbol{\mathscr{L}} \cdot \boldsymbol{\mathscr{L}} = \mathscr{L}^2 = \mathscr{L}_x^2 + \mathscr{L}_y^2 + \mathscr{L}_z^2$$

The scalar quantity $\mathscr{L}$, being the norm of the vector $\boldsymbol{\mathscr{L}}$, is known as the total angular momentum.

Now as is usual in quantum mechanics, from these expressions for the classical angular momenta may be constructed Hermitian operators by leaving the position coordinates unchanged and replacing p_x by $(\hbar/i)(\partial/\partial x)$, and so on. The resulting operators may readily be shown to commute with the Hamiltonian,

$$\mathscr{L}_x H = H\mathscr{L}_x$$
$$\mathscr{L}_y H = H\mathscr{L}_y$$
$$\mathscr{L}_z H = H\mathscr{L}_z$$
$$\mathscr{L}^2 H = H\mathscr{L}^2$$

They do not, however, all commute among themselves, for while $\mathscr{L}^2$ commutes with each of its components, $\mathscr{L}^2\mathscr{L}_x = \mathscr{L}_x\mathscr{L}^2$, and so on, none of the components commutes with any other component:

$$\mathscr{L}_x\mathscr{L}_y \neq \mathscr{L}_y\mathscr{L}_x \ldots$$

Our interpretation of these commutation rules is the following. There exist eigenfunctions of H that are simultaneously eigenfunctions of $\mathscr{L}^2$ and one of its components, say $\mathscr{L}_z$. When the system is described by such an eigenfunction, then the total energy E, the total angular momentum $\mathscr{L}$, and one of the components of angular momentum, say $\mathscr{L}_z$, have definite values. At the same time $\mathscr{L}_x$ and $\mathscr{L}_y$ are not definite. Of course, states can also be found for which E, $\mathscr{L}$, and $\mathscr{L}_x$ are definite while $\mathscr{L}_y$ and $\mathscr{L}_z$ are indefinite. Such eigenfunctions are just as "true" as any other selection satisfying the Schrödinger equation together with its auxiliary conditions. Which set of states will actually be observed in an experiment depends on the experiment performed, for the apparatus used to observe the system perturbs it in such a way as to call forth a preferred set of states. In any case, the following remains true: states (i.e., eigenfunctions of H) exist that also permit simultaneous observation of $\mathscr{L}^2$ and one of the components of $\boldsymbol{\mathscr{L}}$. By convention the latter is usually taken to be $\mathscr{L}_z$, a choice inspired by the fact that in spherical polar coordinates $\mathscr{L}_z$ has a simpler appearance than either $\mathscr{L}_x$ or $\mathscr{L}_y$.

The student is advised at this point to turn back to Figures 1.11, 1.12, and 1.13 to review the geometric pictures he should learn to associate with commuting, symmetric matrices, and their abstract analogs: commuting, Hermitian operators.

PROBLEM

5.2 (1) In Cartesian coordinates construct the angular momentum operators $\mathscr{L}_x$, $\mathscr{L}_y$, $\mathscr{L}_z$ for a single particle.
(2) Show that $\mathscr{L}_x\mathscr{L}_y - \mathscr{L}_y\mathscr{L}_x = i\hbar\mathscr{L}_z$.

5.6 EIGENVALUES OF THE ANGULAR MOMENTUM OPERATORS

While our investigation of the commutation rules for the angular momentum operators and the Hamiltonian has been sufficient to establish the existence of common eigenfunctions for H, $\mathscr{L}^2$, and $\mathscr{L}_z$, we have yet to establish what these eigenfunctions are and what eigenvalues they will give when operated on successively by H, $\mathscr{L}^2$, and $\mathscr{L}_z$. The eigenvalue E of the Hamiltonian operator occurs only in the radial differential equation 5.3, and it is apparent that we shall be unable to establish energy levels for the central field problem without a specific choice of the potential energy $V(r)$. When cast into the notation of spherical polar coordinates, however, the operators $\mathscr{L}^2$ and $\mathscr{L}_z$ prove to be independent of r, so that a knowledge of the angular part of the wave function is sufficient to select those eigenfunctions of H that are simultaneously eigenfunctions of $\mathscr{L}^2$ and $\mathscr{L}_z$. For example, in spherical polar coordinates $\mathscr{L}_z$ is

$$\mathscr{L}_z = \frac{\hbar}{i}\frac{\partial}{\partial\varphi}$$

whence it is not difficult to show that the functions Y_{lm} (equations 5.6) are already eigenfunctions of $\mathscr{L}_z$ with eigenvalue $m\hbar$:

$$\mathscr{L}_z Y_{lm} = \frac{\hbar}{i}\frac{\partial}{\partial \varphi} Y_{lm} = m\hbar Y_{lm}$$

For reasons that will be better understood in Section 6.10, this dependence of the eigenvalues of $\mathscr{L}_z$ on m has given m the name "magnetic quantum number."

A parallel investigation of the operator $\mathscr{L}^2$ shows that the Y_{lm} are also eigenfunctions of $\mathscr{L}^2$ with eigenvalue $l(l + 1)\hbar^2$,

$$\mathscr{L}^2 Y_{lm} = l(l + 1)\hbar^2 Y_{lm}$$

and this relationship has given l the name "angular momentum quantum number."

To summarize, the energy E will in general depend on the angular momentum quantum number l. Corresponding to each l will be $2l + 1$ degenerate states all of which have the same value of the total angular momentum $\mathscr{L} = \sqrt{l(l + 1)}\hbar$. The states will differ in the z component of angular momentum with one state for each of the possibilities

$$-l\hbar, (-l + 1)\hbar, \ldots, 0, \ldots, (l - 1)\hbar, l\hbar$$

5.7 A VECTOR MODEL

This information can be mentally retained and classified by use of a convenient mnemonic device. We imagine the classical angular momentum vector $\mathscr{L} = (\mathscr{L}_x, \mathscr{L}_y, \mathscr{L}_z)$ to be drawn from the origin of a three dimensional vector space. Then for a given value of the quantum number l, $\mathscr{L}$ has a fixed length $\sqrt{l(l + 1)}\hbar$ and for different values of m can thus only assume different orientations in the space. Quantum restrictions limit the z component $\mathscr{L}_z$ to $2l + 1$ distinct possibilities, but simultaneous measurements of the $\mathscr{L}_x$ and $\mathscr{L}_y$ components, while always occurring as integer multiples of $\hbar$, are not predictable for any individual experiment and may have any values consistent with the restrictions on $\mathscr{L}$ and $\mathscr{L}_z$. Diagrams that represent the explicit situation $l = 1$; $m = 1, 0, -1$ are drawn in Figure 5.3. The end of the vector $\mathscr{L}$ always lies on a sphere of radius $\sqrt{1(1 + 1)}\hbar = \sqrt{2}\hbar$. For the $2l + 1 = 3$ values of m, the z component of angular momentum assumes values $\mathscr{L}_z = +\hbar, 0, -\hbar$. The $\mathscr{L}_x$ and $\mathscr{L}_y$ components are not

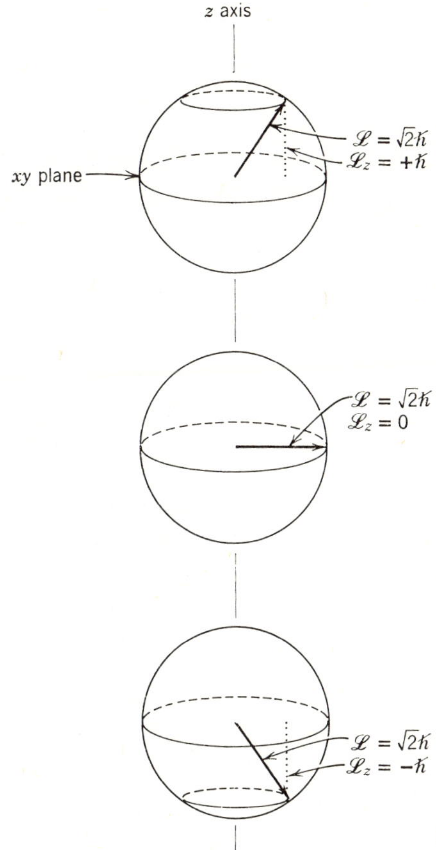

Figure 5.3 Angular momentum states for the central field problem.

further specified, so that the end of the vector may lie at any point along the farther edge of the inscribed cylinder of altitude $\mathscr{L}_z$.

PROBLEMS

5.3 It may be shown that the operator $\mathscr{L}_x^2 + \mathscr{L}_y^2$ commutes with H, $\mathscr{L}^2$, and $\mathscr{L}_z$, so that all four operators may simultaneously have definite eigenvalues—that is, states exist for which all four quantities are definite. Knowing the eigenvalues for $\mathscr{L}^2$ and $\mathscr{L}_z$, calculate the eigenvalues for $\mathscr{L}_x^2 + \mathscr{L}_y^2$.

5.4 Draw vector diagrams corresponding to Figure 5.3 for the case $l = 2$.

5.8 THE REAL ANGULAR FUNCTIONS

While we have demonstrated the functions Y_{lm} to be simultaneously eigenfunctions of both $\mathscr{L}^2$ and $\mathscr{L}_z$, we have yet to investigate the behavior of those linear combinations (5.7) of them that are real when they are operated on by the angular momentum operators. When this is done, Y_x, Y_y, Y_z, and their like prove still to be eigenfunctions of $\mathscr{L}^2$ with eigenvalues $l(l+1)\hbar^2$, but they fail to be eigenfunctions of $\mathscr{L}_z$. For states described by these real angular functions, therefore, the total angular momentum is definite, but we have no idea how the classical angular momentum vector is oriented, even with respect to the component $\mathscr{L}_z$.

However curious these results may seem to the student, I would remind him at this point that quantum mechanics, being a discipline of licensed ignorance, will not supply us with information which it is not within the power of Nature to grant. The curious indeterminacy that is a feature of the phenomenon of degeneracy arises because we are unable to select preferred states (i.e., a preferred basis of eigenfunctions) whenever those states belong to the same value of the energy eigenvalue. It may be that a particular type of experiment will select a preferred group of states to be observed, but a different experiment will more than likely select a totally different set of degenerate eigenfunctions. Whatever the ultimate "truth" concerning the system may be—that is, whatever its description may be when we are not observing it—we are unable to say. On the other hand, these methods do permit us to predict the outcomes of experiments performed on the system, which is as much as the scientist should require for the practical purposes of his work.

5.9 THE HYDROGEN ATOM

An electron moving in the central field of a massive proton constitutes a model for the hydrogen atom.* We assume the Coulomb force law, with potential energy

$$V(r) = -\frac{e^2}{r}$$

* Actually, this is not quite accurate, for the proton, while massive compared to the electron, is not infinitely massive, hence is not strictly fixed to the origin of a central field. To the accuracy needed in chemical calculations, the error is negligible and is in any case eliminated by interpreting the mass of the electron m_e in the Hamiltonian operator as its "reduced mass."[3,4]

in which $-e$ = the charge on the electron,

$$e = 4.803 \times 10^{-10} \text{ abs esu}$$

When this potential energy is substituted into the Schrödinger equation for the central field problem, the equation separates as shown in Section 5.3, and the radial equation 5.3 becomes

$$\frac{1}{r^2}\frac{d}{dr}r^2\frac{dR}{dr} + \frac{2m_e}{\hbar^2}\left(E + \frac{e^2}{r}\right)R - \frac{l(l+1)}{r^2}R = 0 \tag{5.8}$$

in which I have replaced λ by its required value $\lambda = l(l + 1)$, and μ is now identified with m_e, the mass of the electron:

$$m_e = 9.109 \times 10^{-28} \text{ g}$$

It is shown in standard textbooks on quantum mechanics[5] that the solutions to this equation change radically in character depending on whether E is to be considered positive or negative. Because V is everywhere negative, E is positive only when the kinetic energy exceeds in absolute value the potential energy; classically this would imply that the electron is no longer bound to the proton, meaning that the atom is dissociated into its charged constituents. This case is of little interest in chemistry, and we shall not investigate it. For E negative, however, it is found upon investigation that suitable solutions to the equation that satisfy the auxiliary requirements of quantum mechanics exist only if E is restricted to the discrete set of values found long ago by Bohr:

$$E = -\frac{1}{2}\frac{m_e e^4}{n^2\hbar^2}; \qquad n = 1, 2, 3, \ldots$$

$$= -\frac{313.8}{n^2} \quad \text{kcal/mole} \tag{5.9}$$

in which n is another quantum number (the "principal quantum number") and l is hereafter restricted by $l \leq n - 1$.

The student may feel a little cheated by this, for I had promised him that the energy levels would also depend on l. In fact, the energy levels do depend on l for every central field problem other than that of the hydrogen atom. For the Coulomb potential alone do we observe the failure of this rule, so that in this special case not only are all states with the same l and different values of m degenerate, but so are also all states with the same value of n but different values of l.

Radial functions R_{nl} which are associated with an energy level E_n may be expressed in terms of what are known to the mathematician as the associated Laguerre polynomials. I shall serenely pass over a discussion of their general properties, referring the reader to the lists in Appendix III and to more detailed texts.[6–9] Explicit formulas for the first three radial functions are

$$R_{10} = 2a_0^{-3/2} \exp\left(-\frac{r}{a_0}\right)$$

$$R_{20} = \left(\frac{1}{2\sqrt{2}}\right) a_0^{-3/2}\left(2 - \frac{r}{a_0}\right) \exp\left(-\frac{r}{2a_0}\right)$$

$$R_{21} = \left(\frac{1}{2\sqrt{6}}\right) a_0^{-3/2}\left(\frac{r}{a_0}\right) \exp\left(-\frac{r}{2a_0}\right)$$

in which $a_0 = \hbar^2/m_e e^2 = 0.529$ Å known as the Bohr radius has become a standard unit of length in quantum mechanical calculations. The radial functions have been normalized according to the rule

$$\int_0^\infty r^2 R_{nl}^* R_{nl}\, dr = 1$$

Of the formulas listed, the second two belong to a degenerate set, for they have the same value of $n = 2$.

PROBLEM

5.5 Check the orthogonality of the R_{nl} listed above under the rule

$$\int_0^\infty r^2 R_{nl}^* R_{\alpha\beta}\, dr = 0$$

unless both $n = \alpha$ and $l = \beta$.

5.10 DEGENERACY

We can summarize our investigation of the wave mechanical properties of the hydrogen atom as follows. A distinct wave function $\psi_{nlm} = R_{nl}(r) Y_{lm}(\theta, \varphi)$ exists for every distinct selection of quantum numbers n, l, m from the rows

$$n = 1, 2, 3, \ldots$$

$$l = 0, 1, \ldots, n - 1$$

$$m = -l, -l + 1, \ldots, 0, \ldots, l - 1, l$$

Wave functions that have the same n but differ in l and m will belong to degenerate sets. Thus for $n = 1$, there is only the single possibility $l = m = 0$, but for $n = 2$ we have four degenerate wave functions:

$$l = 0; \qquad m = 0$$

$$l = 1; \qquad \begin{cases} m = -1 \\ m = 0 \\ m = 1 \end{cases}$$

For $n = 3$ there are the 9 possibilities:

$$l = 0; \qquad m = 0$$

$$l = 1; \qquad \begin{cases} m = -1 \\ m = 0 \\ m = 1 \end{cases}$$

$$l = 2; \qquad \begin{cases} m = -2 \\ m = -1 \\ m = 0 \\ m = 1 \\ m = 2 \end{cases}$$

By continuing this list the reader will soon perceive that each energy level E_n is n^2 fold degenerate.

A trade jargon that grew up before the development of modern quantum mechanics is still used to describe these degenerate sets of functions. The quantum number l is not specified directly but by a letter code $s, p, d, f, g, \ldots,$ which has the interpretation

$$\begin{array}{llllll} \text{jargon} = s & p & d & f & \ldots \\ l \quad\;\; = 0 & 1 & 2 & 3 & \ldots \end{array}$$

Thus an electron in a state that is described by a wave function whose quantum numbers are $n = 1$, $l = 0$ is a $1s$ electron, an electron in a state whose quantum numbers are $n = 2$, $l = 1$ is a $2p$ electron, and so on. In this notation the hierarchy of energy states is [$1s$]; [$2s$, $2p$]; [$3s$, $3p$, $3d$]; [$4s$, $4p$, $4d$, $4f$]; . . . , where I have bracketed the degenerate sets. If it is desired to specify values of the quantum number m, we write [$1s$]; [$2s$, $2p_{-1}$, $2p_0$, $2p_1$]; [$3s$, $3p_{-1}$, $3p_0$, $3p_1$, $3d_{-2}$, $3d_{-1}$, $3d_0$, $3d_1$, $3d_2$]; . . . , and so forth. Finally, if the real forms of the angular parts of the wave function are used, we write [$1s$], [$2s$, $2px$, $2py$, $2pz$] . . . , with more complicated notations for the states of higher energy.

Because of its later application to problems of chemical bonding in molecules, note the fact that since the energy of the hydrogen atom depends only on n, normalized linear combinations of wave functions that have the same n but different values of l are equally good descriptions of the atom. Thus if we can find four normalized, mutually orthogonal linear combinations of the four wave functions $2s$, $2px$, $2py$, $2pz$, these will serve as well as any other to describe the hydrogen atom in its first excited state. From what I have said before, the implication here is that when the atom is perturbed by the proper experimental observation, such linear combinations may actually become preferred, being in a very real sense the atom's response to our experimental probing.

PROBLEM

5.6 A normalized linear combination of the degenerate $2s$, $2px$, $2py$, $2pz$ wave functions of hydrogen that is of particular interest in chemistry is the one in which all four are mixed in equal proportions:

$$\psi_1 = \tfrac{1}{2}[(2s) + (2px) + (2py) + (2pz)]$$

Remember that only the p orbitals have directional character, so that the directional character of ψ_1 is determined by the sum

$$(2px) + (2py) + (2pz)$$

that is, a concentration of charge occurs along the vector $(1, 1, 1)$ in three dimensional space. Similarly the function

$$\psi_2 = \tfrac{1}{2}[(2s) - (2px) - (2py) + (2pz)]$$

mixes $2s$ and $2p$ in equal proportions, is quantum mechanically orthogonal to ψ_1, and has a maximum charge concentration along the $(-1, -1, 1)$ direction in three dimensional space.

(1) Find two other functions ψ_3 and ψ_4 that mix $2s$ and $2p$ functions in equal proportions and such that ψ_1, ψ_2, ψ_3, ψ_4 are normalized and mutually orthogonal.

(2) Along what directions in three dimensional space do the maximum charge concentrations of ψ_3 and ψ_4 lie? What are the angles in three dimensional space between the directions of maximum charge concentration of ψ_1 and ψ_2, of ψ_3 and ψ_4, and of ψ_1 and ψ_3?

5.11 COMPLETE WAVE FUNCTIONS FOR THE HYDROGEN ATOM

Lists of explicit formulas for the wave functions of hydrogen in real form are given in Appendix III. Three dimensional diagrams of the real functions

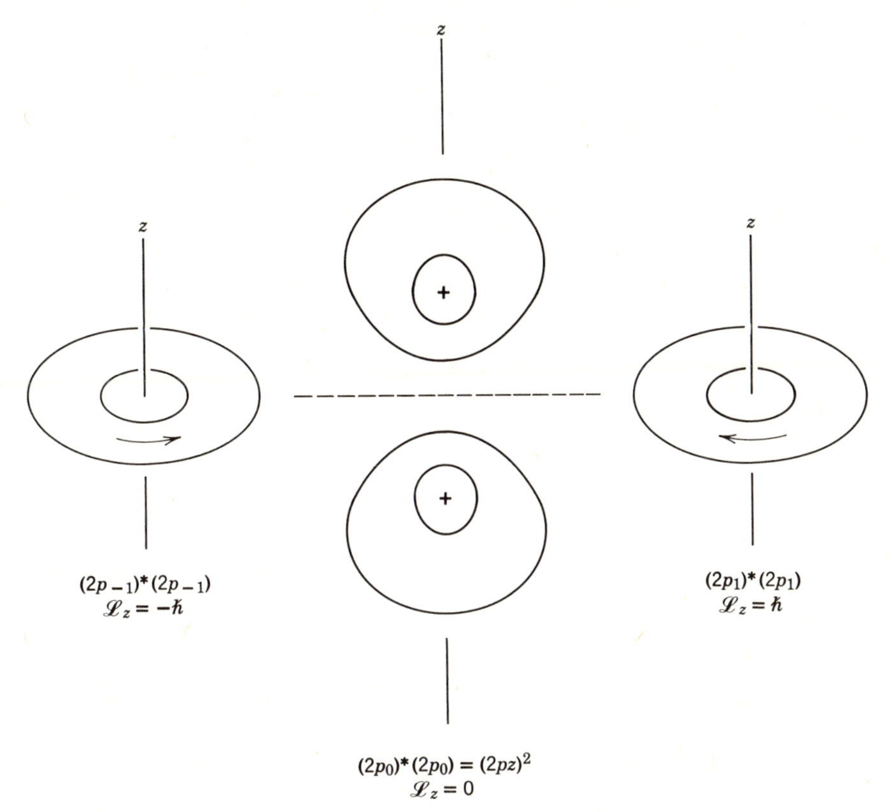

Figure 5.4 Possible first excited states of hydrogen. The z component of angular momentum is definite.

are also available.[10] In counterdistinction to these well known pictures, the reader may be interested in the electron density distributions for $2p$ states implied by the wave functions in their complex form (Figure 5.4). The $2p_{-1}$ state has a nodal line along the z axis and a torus of charge enclosing it. The $\mathscr{L}_z$ component of angular momentum is definite in this state and implies a classical circulation of negative charge in the direction indicated. The diagram for the $2p_1$ state is similar except that the classical circulation of negative charge is in the opposite direction. The $2p_0$ state is identical with the $2pz$ state for which $\mathscr{L}_z = 0$. Except for orientation in space, the contours of $(2pz)^2$ are calculated by squaring those of Figure 5.2.

REFERENCES

1. W. E. Byerly, *Fourier Series and Spherical Harmonics*, Dover Reprint, New York, 1959, pp. 198–199.

2. H. Eyring, J. Walter, and G. E. Kimball, *Quantum Chemistry*, John Wiley and Sons, Inc., New York, 1944, pp. 52–59.
3. H. Eyring, J. Walter, and G. E. Kimball, *ibid.*, pp. 80–81.
4. L. Pauling and E. B. Wilson, Jr., *Introduction to Quantum Mechanics*, McGraw-Hill Inc., New York, 1935, p. 114.
5. H. Eyring, J. Walter, and G. E. Kimball, *loc. cit.*, pp. 90–91.
6. H. Eyring, J. Walter, and G. E. Kimball, *loc. cit.*, pp. 63–67.
7. R. Courant and D. Hilbert, *Methods of Mathematical Physics*, Vol. 1, Interscience, New York, 1953, pp. 93–95.
8. P. M. Morse and H. Feshbach, *Methods of Theoretical Physics*, Vol. 1, McGraw-Hill, Inc., New York, 1953, pp. 784–785.
9. F. L. Pilar, *Elementary Quantum Chemistry*, McGraw-Hill, Inc., New York, 1968, pp. 154–156.
10. C. A. Coulson, *Valence*, Oxford Press, 1961, pp. 24–25, 43.

Chapter 6

PERTURBATION THEORY

6.1 THE NEED FOR APPROXIMATE METHODS

The problems discussed so far, whether stated in terms of a finite dimensional vector space, as were the Hückel molecular orbitals of Chapter 2, or in terms of a function space, as was the central field problem of Chapter 5, have all been exactly solvable. An exactly solvable problem in quantum mechanics is rarely encountered, and when one is encountered, it may confidently be assumed that the eager beavers of the 1920s have probably already solved it. There remain a host of interesting insights into quantum mechanics that can be obtained by way of approximate methods, and therein lies much of the research effort that is being put into the subject nowadays. As has so far been customary in this text, I shall first demonstrate the application of our methods to finite dimensional vector spaces before broadening them to the conceptually more difficult function spaces.

6.2 A LEMMA ON SYMMETRIC MATRICES

Let us consider a real vector space and a symmetric matrix $K = K^T$. Then for *any* two vectors $\mathbf{x}$ and $\mathbf{y}$ the lemma states that

$$\mathbf{y}^T K\mathbf{x} = \mathbf{x}^T K\mathbf{y}$$

The proof is very simple. Both sides of the equation are scalar quantities and must therefore be indifferent to transposition. Let us take the transpose of the right hand side, rearranging the order of the factors according to the rules 1.2:

$$\mathbf{x}^T K\mathbf{y} = (\mathbf{x}^T K\mathbf{y})^T = \mathbf{y}^T K^T\mathbf{x}$$

But $K = K^T$, so that

$$\mathbf{x}^T K\mathbf{y} = \mathbf{y}^T K\mathbf{x}$$

as was to be proved. This lemma has in fact already been used in Section 1.22. In establishing our theorem on the orthogonality of those eigenvectors of a real, symmetric matrix that belong to different eigenvalues. There is, as usual, an analog to this lemma for Hermitian operators in a function space. In fact, a statement of the lemma is traditionally taken in advanced texts[1,2] on quantum mechanics to be the defining property of an Hermitian operator.

6.3 THE PERTURBATION ARGUMENT

Many approximate calculations may be approached from the point of view that the problem in hand differs only slightly from a simpler one whose exact solution is already known. Thus suppose that we have an eigenvalue problem in a finite dimensional vector space

$$H\psi = E\psi \tag{6.1}$$

and that H is the sum of two symmetric matrices $H = H^0 + \varepsilon H^1$, where ε is a scalar and the product εH^1 may be assumed to be small. The solutions to the zero order eigenvalue problem

$$H^0\psi^0 = E^0\psi^0$$

are presumed to be known, and we may therefore imagine that for εH^1 sufficiently small, the eigenvectors ψ and eigenvalues E of the perturbed problem should not differ markedly from ψ^0 and E^0.

Let us suppose that a particular $E_k^{\,0}$ is nondegenerate with an eigenvector $\psi_k^{\,0}$ and consider the effect of the perturbation εH^1 upon this state. Expand the unknown eigenvalue as a power series in ε

$$E = E_k^{\,0} + \varepsilon E_k^{\,1} + \mathcal{O}(\varepsilon^2) + \cdots \tag{6.2}$$

in which $\mathcal{O}(\varepsilon^2)$ means that all subsequent terms are multiplied by powers of ε greater than or equal to 2. In equation 6.2 we are to determine $E_k^{\,1}$ and the coefficients of higher powers of ε. A similar expansion for the eigenvector is

$$\psi = \psi_k^{\,0} + \varepsilon\psi_k^{\,1} + \mathcal{O}(\varepsilon^2) + \cdots \tag{6.3}$$

in which $\psi_k^{\,1}$, the vector coefficient of ε, is to be determined.

When these expansions are substituted into the original problem (6.1),

$$(H^0 + \varepsilon H^1)(\psi_k^{\,0} + \varepsilon\psi_k^{\,1} + \cdots) = (E_k^{\,0} + \varepsilon E_k^{\,1} + \cdots)(\psi_k^{\,0} + \varepsilon\psi_k^{\,1} + \cdots)$$

or

$$H^0\psi_k^{\,0} + \varepsilon[H^0\psi_k^{\,1} + H^1\psi_k^{\,0}] + \cdots = E_k^{\,0}\psi_k^{\,0} + \varepsilon[E_k^{\,0}\psi_k^{\,1} + E_k^{\,1}\psi_k^{\,0}] + \cdots$$

and on each side of the equation we equate coefficients of like powers of ε, we have for the zeroth order term

$$H^0\psi_k^0 = E_k^0\psi_k^0$$

which is satisfied automatically. From the coefficient of ε, we have a first order correction to the eigenvalue problem

$$H^0\psi_k^1 + H^1\psi_k^0 = E_k^0\psi_k^1 + E_k^1\psi_k^0 \tag{6.4}$$

from which both E_k^1 and ψ_k^1 are to be determined.

The methods of vector algebra enable us to handle the details efficiently. Because the eigenvectors of H^0 form a basis in the vector space of n dimensions, we can expand the unknown vector ψ_k^1 in terms of the ψ_j^0:

$$\psi_k^1 = \sum_j c_{jk}\psi_j^0 \tag{6.5}$$

in which the c_{jk} are unknown. Similarly, we expand

$$H^1\psi_k^0 = \sum_j b_{jk}\psi_j^0 \tag{6.6}$$

in which the b_{jk} are known coefficients to be calculated from

$$b_{jk} = (\psi_j^0)^T H^1\psi_k^0; \qquad j, k = 1, 2, \ldots, n$$

(If you do not remember how to do this, see Section 1.15.) Note that because of the lemma of Section 6.2, the b_{jk} are the components of an n dimensional, symmetric matrix $b_{jk} = b_{kj}$.

Expansions 6.5 and 6.6 are now to be substituted into equation 6.4 for the first order correction, and we let H^0 operate on ψ_k^1,

$$H^0\psi_k^1 = H^0\sum_j c_{jk}\psi_j^0 = \sum_j c_{jk}E_j^0\psi_j^0$$

so that (6.4) becomes

$$\sum_j c_{jk}E_j^0\psi_j^0 + \sum_j b_{jk}\psi_j^0 = E_k^0\sum_j c_{jk}\psi_j^0 + E_k^1\psi_k^0 \tag{6.7}$$

Now according to the theorem of Section 1.13, because the ψ_j^0 constitute a set of n linearly independent vectors in the vector space of n dimensions, equation 6.7 can be satisfied only if the coefficients of identical vectors on either side of equation 6.7 are separately equal. From the coefficients of ψ_k^0 we have

$$b_{kk} = E_k^1$$

thus determining our first order correction to the eigenvalue. From the coefficients of the other ψ_j^0, $j \neq k$, we obtain

$$c_{jk} = \frac{b_{jk}}{E_k^0 - E_j^0}, \qquad j \neq k$$

which are the quantities required for a first order correction to the eigenvector. To the first order of small quantities we therefore have as a result of our perturbation calculation an (unnormalized) eigenvector

$$\psi = \psi_k^{\,0} + \varepsilon \sum_{j \neq k} b_{jk} \psi_j^{\,0} / (E_k^{\,0} - E_j^{\,0}) + \mathcal{O}(\varepsilon^2) \tag{6.8}$$

with its associated eigenvalue

$$E = E_k^{\,0} + \varepsilon (\psi_k^{\,0})^T H^1 \psi_k^{\,0} + \mathcal{O}(\varepsilon^2) \tag{6.9}$$

6.4 SECOND ORDER PERTURBATION THEORY

The first order correction to the eigenvalue problem was obtained from the coefficient of ε in the expansions of E and ψ into power series in ε. In a similar way, higher order corrections can be obtained from the coefficients of higher powers of ε. Such higher order calculations are occasionally carried out, but we shall have no need for them in this text.

6.5 HÜCKEL MOLECULAR ORBITALS FOR FORMALDEHYDE

Let us carry out a complete, first order perturbation calculation for the formaldehyde molecule $H_2C = O$ considered as a perturbation of ethylene in that one of the carbons is replaced by oxygen. The effect of such a substitution on our matrix H will be to make more negative the diagonal matrix element

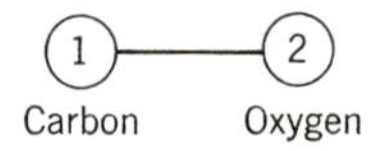

H_{22} which corresponds to the oxygen atom; for oxygen has a greater nuclear charge than carbon, and removing a $2p$ electron from an isolated oxygen atom should require more energy than the corresponding process for carbon. The availability in oxygen of a lower lying energy state for a valence electron than is the case for carbon is described by saying that O is more *electronegative* than C. I shall therefore choose for the ionization energy of oxygen a matrix element $H_{22} = \alpha + \beta$, where α and β have their old meanings. The reasoning behind this particular choice is frankly empirical, and for it the reader is directed elsewhere.[3]

The H matrix whose eigenvalues and eigenvectors we seek is then

$$H = \begin{pmatrix} \alpha & \beta \\ \beta & \alpha + \beta \end{pmatrix}$$

and the reader will note that in terms of the notation of Section 6.3 H is the sum $H^0 + \varepsilon H^1$ where $\varepsilon = 1$,

$$H^0 = \begin{pmatrix} \alpha & \beta \\ \beta & \alpha \end{pmatrix}$$

is the unperturbed ethylene matrix, and

$$H^1 = \begin{pmatrix} 0 & 0 \\ 0 & \beta \end{pmatrix}$$

is the perturbation introduced by the substitution of the oxygen atom for carbon. From Section 2.4, the unperturbed eigenvalues and eigenvectors of H^0 are available. They are

$$\begin{aligned} E_1{}^0 &= \alpha + \beta; \qquad \psi_1{}^0 = \left(\frac{1}{\sqrt{2}}, \frac{1}{\sqrt{2}}\right) \\ E_2{}^0 &= \alpha - \beta; \qquad \psi_2{}^0 = \left(\frac{1}{\sqrt{2}}, -\frac{1}{\sqrt{2}}\right) \end{aligned} \tag{6.10}$$

The coefficients b_{jk} are now calculated from the zero order eigenvectors

$$b_{jk} = (\psi_j{}^0)^T H^1 \psi_k{}^0 = b_{kj}$$

whence

$$b_{11} = \left(\frac{1}{\sqrt{2}}, \frac{1}{\sqrt{2}}\right) \begin{pmatrix} 0 & 0 \\ 0 & \beta \end{pmatrix} \begin{pmatrix} \frac{1}{\sqrt{2}} \\ \frac{1}{\sqrt{2}} \end{pmatrix} = \tfrac{1}{2}\beta$$

$$b_{22} = \left(\frac{1}{\sqrt{2}}, -\frac{1}{\sqrt{2}}\right) \begin{pmatrix} 0 & 0 \\ 0 & \beta \end{pmatrix} \begin{pmatrix} \frac{1}{\sqrt{2}} \\ -\frac{1}{\sqrt{2}} \end{pmatrix} = \tfrac{1}{2}\beta$$

$$b_{12} = b_{21} = \left(\frac{1}{\sqrt{2}}, -\frac{1}{\sqrt{2}}\right) \begin{pmatrix} 0 & 0 \\ 0 & \beta \end{pmatrix} \begin{pmatrix} \frac{1}{\sqrt{2}} \\ \frac{1}{\sqrt{2}} \end{pmatrix} = -\tfrac{1}{2}\beta$$

For the ground state we have therefore from (6.2) a perturbed eigenvalue

$$E_1 = E_1{}^0 + \varepsilon b_{11} = \alpha + (1.5)\beta$$

and from (6.3) a perturbed (unnormalized) ground state eigenvector

$$\psi_1 = \psi_1^0 + \frac{b_{21}}{E_1^0 - E_2^0}\psi_2^0 = \psi_1^0 - \frac{\frac{1}{2}\beta}{2\beta}\psi_2^0 = \psi_1^0 - \tfrac{1}{4}\psi_2^0 \qquad (6.11)$$

To normalize ψ_1, we have only to divide through the coefficients on the right of (6.11) by the square root of the sum of their squares $[(1)^2 + (\frac{1}{4})^2]^{1/2} = 1.031$, whence

$$\psi_1 = 0.970\psi_1^0 - 0.243\psi_2^0$$

or, substituting from (6.10),

$$\psi_1 = (0.514, 0.858)$$

The treatment of the second eigenvalue and eigenvector of formaldehyde is similar, and there results

$$E_2 = \alpha - (0.5)\beta$$

$$\psi_2 = (0.858, -0.514)$$

Now, according to the exclusion principle the two bonding electrons that make up the π bond in formaldehyde can both be accommodated in the ground state. Their total energy will be $2E_1 = 2\alpha + 3\beta$. If the bond is broken, the energies of the electrons located on their separate atoms are α (for carbon) plus $\alpha + \beta$ (for oxygen) $= 2\alpha + \beta$. It follows that the bond energy is $2\alpha + 3\beta - (2\alpha + \beta) = 2\beta$ just as for ethylene.

For the charge distribution in the molecule we have on the carbon the contribution of both electrons in ψ_1

$$2(0.514)^2 = 0.528 \text{ (carbon)}$$

and correspondingly for oxygen

$$2(0.858)^2 = 1.472 \text{ (oxygen)}$$

The more electronegative oxygen atom has thus a greater concentration of electronic charge about it. When we subtract these figures from the residual charge +1 of each nucleus we obtain a picture of the molecular charge distribution.

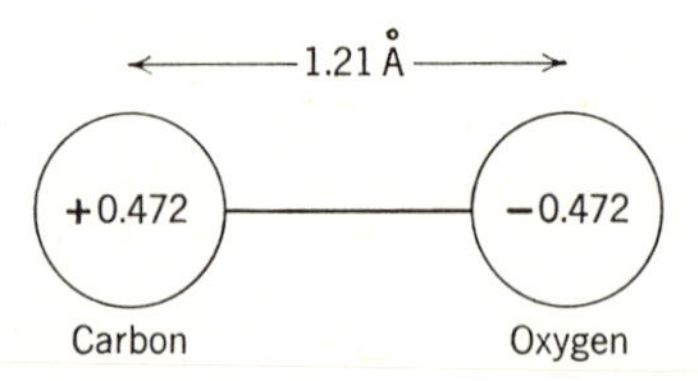

I have also written in the experimental carbon-oxygen interatomic bond distance 1.21 Å. The dipole moment μ of this molecule is defined to be the product of the charge concentration about the positive pole times the interatomic distance:

$$\begin{aligned}\mu &= (0.472)e(1.21 \times 10^{-8}\text{ cm})\\ &= (0.472)(4.803 \times 10^{-10}\text{ esu})(1.21 \times 10^{-8}\text{ cm})\\ &= 2.74\text{ debye units}\end{aligned}$$

The best experimental values for the dipole moment of formaldehyde range from 2.29 to 2.34 debye.[4]

When we find exactly the eigenvalues and eigenvectors of

$$H = \begin{pmatrix} \alpha & \beta \\ \beta & \alpha + \beta \end{pmatrix}$$

the results from Problem 2.6 prove to be

$$\begin{aligned} E_1 &= \alpha + 1.618\beta; \qquad \psi_1 = (0.526, 0.851)\\ E_2 &= \alpha - 0.618\beta; \qquad \psi_2 = (0.851, -0.526)\end{aligned}$$

These are to be compared with the approximate results obtained above by a first order perturbation calculation. For the energy of the double bond I find 2.236β (i.e., stronger than ethylene) and for the dipole moment 2.60 debye.

6.6 DEGENERATE FIRST ORDER PERTURBATION THEORY

The foregoing calculation for a nondegenerate matrix H^0 must be modified when some of the eigenvalues of the unperturbed matrix are degenerate. The basic difficulty is that we do not know precisely the zero order eigenvectors ψ_j^0 that correspond to the degenerate level. We know them only to within some arbitrary linear combination of an arbitrarily chosen degenerate set of basis vectors. To illustrate this difficulty and our method of overcoming it, I shall suppose that ψ_1^0 and ψ_2^0 both belong to a doubly degenerate eigenvalue E^0 and that the other eigenvectors are assigned to other eigenvalues. Our problem is the same: given $H = H^0 + \varepsilon H^1$, find the eigenvalues and eigenvectors of H correct to the first order of powers of ε:

$$(H^0 + \varepsilon H^1)\psi = E\psi \tag{6.12}$$

As before, I expand

$$E = E^0 + \varepsilon E^1 + \cdots$$

$$\psi = p\psi_1^0 + q\psi_2^0 + \varepsilon\psi^1 + \cdots$$

in which the coefficients p and q are unknown beyond the fact that normalization of the zeroth order eigenvectors requires $p^2 + q^2 = 1$. (Why?) When these expansions are substituted into (6.12) and on each side of (6.12) the coefficients of like powers of ε are collected and equated to each other, we have for the first order perturbation

$$H^0\psi^1 + H^1(p\psi_1{}^0 + q\psi_2{}^0) = E^0\psi^1 + E^1(p\psi_1{}^0 + q\psi_2{}^0) \tag{6.13}$$

from which ψ^1, E^1, p, and q are to be determined.

Expand

$$\psi^1 = \sum_j c_j\psi_j{}^0$$

so that

$$H^0\psi^1 = \sum_j c_jE_j{}^0\psi_j{}^0$$

with the c's as yet unknown. Similarly expand

$$H^1\psi_1{}^0 = b_{11}\psi_1{}^0 + b_{21}\psi_2{}^0 + \sum_{j=3}^{n} b_{j1}\psi_j{}^0$$

$$H^1\psi_2{}^0 = b_{12}\psi_1{}^0 + b_{22}\psi_2{}^0 + \sum_{j=3}^{n} b_{j2}\psi_j{}^0$$

in which the b_{jk} are calculable from

$$b_{jk} = (\psi_j{}^0)^T H^1\psi_k{}^0 = b_{kj}$$

and I have explicitly written out those terms in the expansion whose eigenvectors are associated with the degenerate state. When these expansions are substituted into the first order perturbation equation (6.13), both sides of the result are expressed in terms of expansions in the orthonormal set $\psi_j{}^0$, and it follows that the coefficients of corresponding terms must be equal. From the coefficients of the degenerate set $\psi_1{}^0$, $\psi_2{}^0$ we obtain in this way the pair of equations

$$pb_{11} + qb_{12} = E^1p$$
$$pb_{21} + qb_{22} = E^1q$$

or

$$(b_{11} - E^1)p + b_{12}q = 0$$
$$b_{21}p + (b_{22} - E^1)q = 0$$

Equations of this type are by now familiar. Solutions for p and q will exist only if the determinant of coefficients vanishes.

$$\begin{vmatrix} b_{11} - E^1 & b_{12} \\ b_{21} & b_{22} - E^1 \end{vmatrix} = 0 \tag{6.14}$$

and the eigenvalues of this secular equation give a pair of allowed values for E^1 together with a pair of associated eigenvectors whose components are the coefficients p and q required to fix definitely our zeroth order eigenvectors for the degenerate state.

We have by no means finished with our perturbation calculation, for we have yet to find the coefficients c_j that will define ψ^1; yet it is worthwhile to pause at this point and ponder the significance of what we have learned concerning the effect of a perturbation on a degenerate state. In the general case, the secular equation 6.14 will furnish us with two distinct roots E_1^1 and E_2^1. The originally doubly degenerate energy level E^0 will thus be split by the perturbation into two distinct levels, $E^0 + \varepsilon E_1^1$ and $E^0 + \varepsilon E_2^1$. Corresponding to each of these new levels is a distinct eigenvector expressed as a power series expansion in ε:

$$\psi_1 = p_1\psi_1^0 + q_1\psi_2^0 + \mathcal{O}(\varepsilon) + \cdots$$

$$\psi_2 = p_2\psi_1^0 + q_2\psi_2^0 + \mathcal{O}(\varepsilon) + \cdots$$

where I have written down only the leading or zeroth order terms. The coefficients (p_1, q_1) and (p_2, q_2) are the components of the normalized eigenvectors of the two dimensional secular equation 6.14. Even if we carry the perturbation calculation no further than this, we have gained considerable information, for we see that a perturbation can act to remove a degeneracy, and that the removal of this degeneracy enables us to resolve the annoying indeterminacy connected with the eigenvectors of a degenerate state: even the zeroth order eigenvectors of the perturbed problem become perfectly definite, and we are no longer plagued with the irritating necessity of admitting that we cannot assign a unique eigenvector to a particular level. Of course, it could turn out that the eigenvalues E_1^1 and E_2^1 derived from (6.14) are the same, and in this case the degeneracy would not be removed nor would the zero order eigenvectors become definite.

This application of perturbation theory to the splitting of the energy levels of a degenerate state and the correct assignment of zero order eigenvectors is by all odds the most important application of it which we shall make to problems in quantum chemistry. Instead of continuing our analysis to complete the calculation of the c_j and thus developing the eigenvectors to first order accuracy, I shall devote the remaining space to reviewing our recipe for handling the degeneracy problem as far as we have taken it. The results will be split, degenerate energy levels calculated to first order accuracy and their associated eigenvectors written down only to zero order accuracy.

We must first identify the degenerate energy level of the unperturbed problem and *any* set of normalized, mutually orthogonal eigenvectors which completely span the degenerate subspace. Suppose the level E^0 to be n fold degenerate and let the unperturbed eigenvectors be $\psi_1^0, \psi_2^0, \ldots, \psi_n^0$.

Construct the n dimensional, symmetric matrix B,

$$b_{ij} = (\psi_i^0)^T H^1 \psi_j^0$$

and find its eigenvalues and normalized eigenvectors. Correct to first order in ε, the split energy levels are

$$E = E^0 + \varepsilon E_k^1$$

in which the E_k^1 are the eigenvalues of B. The originally degenerate level will thus be split into as many branches as there are distinct eigenvalues of B. Finally, if $(p_{k1}, p_{k2}, \ldots, p_{kn})$ is an eigenvector of B with eigenvalue E_k^1, then the zero order eigenvector of H corresponding to an energy level $E = E^0 + \varepsilon E_k^1$ is

$$\psi_k = p_{k1}\psi_1^0 + p_{k2}\psi_2^0 + \cdots + p_{kn}\psi_n^0$$

6.7 HÜCKEL MOLECULAR ORBITALS FOR PYRIDINE

Let us examine the pyridine molecule considered as a perturbation of benzene by replacing one of the ring carbons with nitrogen. As with formaldehyde, the effect which this substitution will have on H is to make more negative that diagonal element of H which corresponds to the energy of an electron on the heteroatom. Designating atom 4 of Figure 6.1 as nitrogen, I set $H_{44} = \alpha + \frac{1}{2}\beta$. As with formaldehyde, this choice is governed by empirical considerations, but when compared with the choice $\alpha + \beta$ for oxygen in formaldehyde, it makes good chemical sense, for nitrogen lies between carbon and oxygen in electronegativity.

The matrix H is thus the sum $H = H^0 + \varepsilon H^1$ in which $\varepsilon = \frac{1}{2}$, H^0 is the

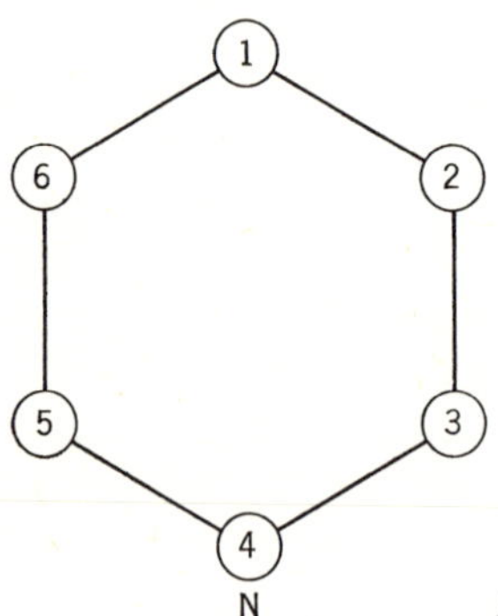

Figure 6.1 Hückel diagram for pyridine.

unperturbed benzene matrix (Section 2.6), and

$$H^1 = \begin{pmatrix} 0 & 0 & 0 & 0 & 0 & 0 \\ 0 & 0 & 0 & 0 & 0 & 0 \\ 0 & 0 & 0 & 0 & 0 & 0 \\ 0 & 0 & 0 & \beta & 0 & 0 \\ 0 & 0 & 0 & 0 & 0 & 0 \\ 0 & 0 & 0 & 0 & 0 & 0 \end{pmatrix}$$

There are two doubly degenerate levels in benzene, and I shall treat the lowest lying one first, using the molecular orbitals ψ_2^0 and ψ_3^0 listed among equations 2.4. We set up the two dimensional matrix B:

$$b_{22} = (\psi_2^0)^T H^1 \psi_2^0 = 0$$
$$b_{33} = (\psi_3^0)^T H^1 \psi_3^0 = \tfrac{1}{3}\beta$$
$$b_{23} = (\psi_2^0)^T H^1 \psi_3^0 = b_{32} = 0$$

The secular equation is

$$\begin{vmatrix} 0 - E^1 & 0 \\ 0 & \tfrac{1}{3}\beta - E^1 \end{vmatrix} = 0$$

whose eigenvalues and normalized eigenvectors are readily found to be

$$E_2^1 = 0; \qquad (1, 0)$$
$$E_3^1 = \tfrac{1}{3}\beta; \qquad (0, 1)$$

It follows that the originally degenerate benzene level $E^0 = \alpha + \beta$ is split in pyridine into two states, the first of which to first order accuracy is unchanged,

$$E_2 = E^0 + \varepsilon E_2^1 = \alpha + \beta$$

and the second of which is lowered by the (negative) amount $\beta/6$,

$$E_3 = E^0 + \varepsilon E_3^1 = E^0 + (\tfrac{1}{2})(\tfrac{1}{3}\beta) = \alpha + \tfrac{7}{6}\beta$$

The correct *zero order* molecular orbitals corresponding to this splitting are

$$\psi_2 = (1)\psi_2^0 + (0)\psi_3^0 = \left(\frac{1}{\sqrt{4}}\right)(0, 1, 1, 0, -1, -1)$$

$$\psi_3 = (0)\psi_2^0 + (1)\psi_3^0 = \left(\frac{1}{\sqrt{12}}\right)(2, 1, -1, -2, -1, 1)$$

meaning that by good fortune we had picked as our original basis a set of degenerate orbitals that was already correctly assigned with regard to the perturbation of the ring by a nitrogen atom at position 4.

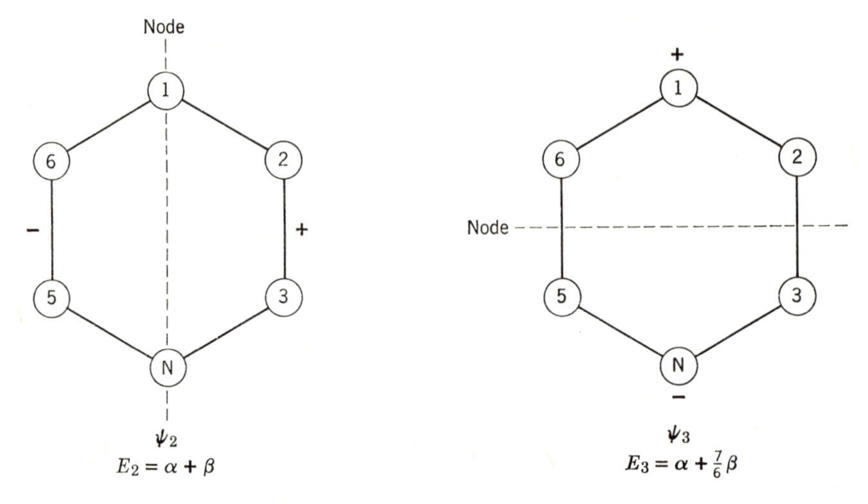

Figure 6.2 Zero order molecular orbitals for pyridine.

It is instructive to note in Figure 6.2 the position of the nodal lines for these zero order wave functions. The nitrogen atom at position 4 lowers the energy of ψ_3 by an amount $\beta/6$, but to first order accuracy it does not change the energy of ψ_2.

6.8 THE SECOND DEGENERATE STATE

Our molecular orbital treatment of benzene showed an energy level at $E^0 = \alpha - \beta$ to which two molecular orbitals $\psi_4{}^0$ and $\psi_5{}^0$ were assigned. Let us apply our perturbation method to this level, using the zero order orbitals listed among equations 2.4. Our perturbation matrix elements are

$$b_{44} = (\psi_4{}^0)^T H^1 \psi_4{}^0 = \tfrac{1}{4}\beta$$
$$b_{55} = (\psi_5{}^0)^T H^1 \psi_5{}^0 = \tfrac{1}{12}\beta$$

$$b_{45} = (\psi_4{}^0)^T H^1 \psi_5{}^0 = b_{54} = \left(\frac{1}{2\sqrt{12}}\right)\beta$$

whence

$$\begin{vmatrix} \dfrac{\beta}{4} - E^1 & \dfrac{\beta}{2\sqrt{12}} \\ \dfrac{\beta}{2\sqrt{12}} & \dfrac{\beta}{12} - E^1 \end{vmatrix} = 0$$

Eigenvalues and normalized eigenvectors are

$$E_4^{\,1} = 0; \qquad \left(-\frac{1}{2}, \frac{\sqrt{3}}{2}\right)$$

$$E_5^{\,1} = \tfrac{1}{3}\beta; \qquad \left(\frac{\sqrt{3}}{2}, \frac{1}{2}\right)$$

so that correct zero order molecular orbitals and their first order energy levels are

$$E_4 = E^0 + \varepsilon E_4^{\,1} = \alpha - \beta$$

$$\psi_4 = -\tfrac{1}{2}\psi_4^{\,0} + \left(\frac{\sqrt{3}}{2}\right)\psi_5^{\,0}$$

$$= \left(\frac{1}{\sqrt{4}}\right)(0, 1, -1, 0, 1, -1)$$

$$E_5 = E^0 + \varepsilon E_5^{\,1} = \alpha - \tfrac{5}{6}\beta$$

$$\psi_5 = \left(\frac{\sqrt{3}}{2}\right)\psi_4^{\,0} + \tfrac{1}{2}\psi_5^{\,0}$$

$$= \left(\frac{1}{\sqrt{12}}\right)(2, -1, -1, 2, -1, -1)$$

Nodal lines for these zero order assignments are sketched in Figure 6.3. The observation should not be lost upon the student that the perturbing effect of the nitrogen atom is to locate definitely in the molecule the positions of the nodal lines of the originally degenerate molecular orbitals of benzene, and

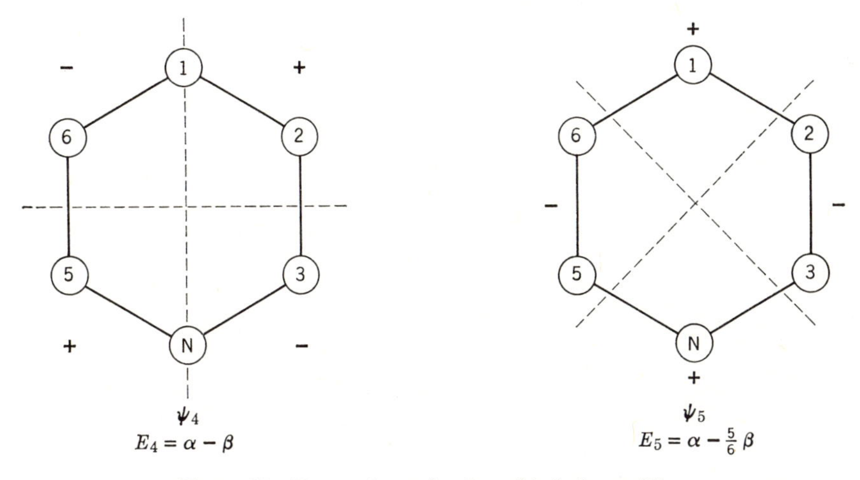

Figure 6.3 Zero order molecular orbitals for pyridine.

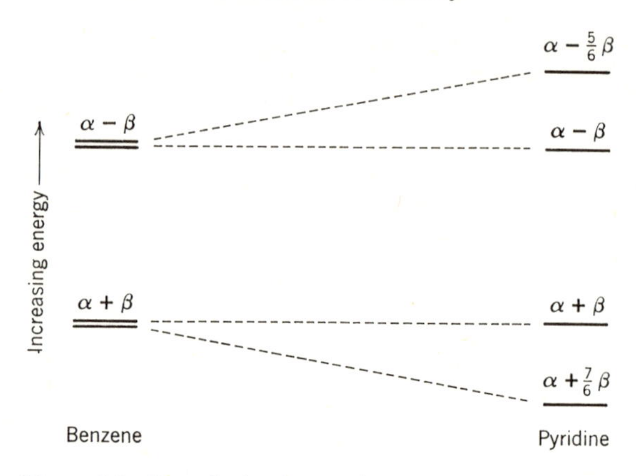

Figure 6.4 Perturbation energy level diagram for pyridine.

that such nodal lines assume positions symmetrically disposed with respect to the nitrogen atom. Finally, we can summarize our observations of the splitting of the degenerate energy states by Figure 6.4.

PROBLEM

6.1 A suitable set of molecular orbitals for the first degenerate level of benzene is

$$\psi_2^0 = \left(\frac{1}{\sqrt{4}}\right)(1, 1, 0, -1, -1, 0)$$

$$\psi_3^0 = \left(\frac{1}{\sqrt{12}}\right)(1, -1, -2, -1, 1, 2)$$

each of which has an unperturbed energy $E^0 = \alpha + \beta$. Calculate to the first order the new energy levels derived from E^0 when benzene is converted to pyridine, and assign to each new level a zero order molecular orbital. Compare your results with those of Section 6.7.

6.9 PERTURBATION THEORY IN FUNCTION SPACES

The foregoing analysis for finite dimensional vector spaces generalizes immediately to the infinite dimensional function space. Those features of this theory that we need in our work are summarized here.

I assume that the Hamiltonian operator H for a given problem can be decomposed into two terms $H = H^0 + \varepsilon H^1$ in which the Schrödinger equation for the simpler problem

$$H^0\psi^0 = E^0\psi^0$$

can be exactly solved and εH^1 may be presumed small. If when $\varepsilon = 0$ a given level E_k^0 is nondegenerate, correct to first order of powers of ε the perturbed energy level is

$$E_k = E_k^0 + \varepsilon b_{kk}$$

where

$$b_{kk} = \int (\psi_k^0)^* H^1 \psi_k^0 \, d\tau$$

The perturbed wave function may also be expanded in powers of ε, but I shall not trouble to do so. Correct to zero order of powers of ε it is simply

$$\psi_k = \psi_k^0$$

If the unperturbed energy level E^0 is n fold degenerate, the perturbed energy levels will in general show a splitting of the original level into different branches or "terms," as they are usually called. Correct to first order of powers of ε, each term has an energy

$$E_k = E^0 + \varepsilon E_k^1$$

where the E_k^1 are the eigenvalues of the n dimensional, symmetric matrix B whose elements are the integrals

$$b_{jk} = \int (\psi_j^0)^* H^1 \psi_k^0 \, d\tau \tag{6.15}$$

Zero order wave functions that are to be assigned to these terms are linear combinations of all the n unperturbed wave functions ψ_j^0 which were associated with the original, unperturbed, degenerate state.

$$\psi_k = p_{k1}\psi_1^0 + p_{k2}\psi_2^0 + \cdots + p_{kn}\psi_n^0$$

The coefficients p_{kj} in this expansion are the components of a normalized eigenvector of B associated with an eigenvalue E_k^1.

The importance of this theory is derived not so much from the quantitative results it yields as from the qualitative insights it gives into problems of atomic and molecular structure. We shall see in Chapter 8 how when combined with the Pauli exclusion principle it enables us to build up the periodic system of the elements in a consistent and logical way. Finally, we shall use it to justify the qualitative picture that the modern chemist has of the chemical bond.

6.10 ZEEMAN EFFECT FOR THE TWO DIMENSIONAL, ISOTROPIC HARMONIC OSCILLATOR

If an atom is placed in a magnetic field, many of the lines of its emission spectrum are observed to split into a set of fine, closely spaced lines, the width of the spacing being proportional to the strength of the externally imposed field (Zeeman effect). We shall examine the theory of this phenomenon for the two dimensional, isotropic, harmonic oscillator of Section 4.9, assuming that the oscillating particle is electrically charged so that the classical result of its orbital motion can be pictured as an electric current flowing in a circuit. Perpendicular to the plane of the circuit is a magnetic dipole, a vector quantity proportional to the angular momentum vector of the particle. For the two dimensional oscillator there is only one component $\mathscr{L}_z$ of angular momentum, and if the oscillator is placed in an external magnetic field of strength $\mathscr{B}$, then the classical particle will take on an additional potential energy $\varepsilon\mathscr{L}_z$ due to the coupling of its magnetic dipole with the external field. The quantity ε is proportional to the strength of the external field and is given explicitly from electromagnetic theory as $\varepsilon = q\mathscr{B}/2\mu c$ in which q is the charge on the oscillator, μ its mass, and c the velocity of light.

These remarks can be translated into quantum mechanical language if we add to the oscillator Hamiltonian on the left of equation 4.3 an operator $\varepsilon H^1 = \varepsilon\mathscr{L}_z = \varepsilon(\hbar/i)(\partial/\partial\varphi)$ and treat this term as a perturbation.

To be specific, let us examine the effect of an external magnetic field upon the oscillator when in the absence of the field it would be in its doubly degenerate first excited state. As zero order wave functions (or as a basis) for the perturbation calculation I choose equations 4.7:

$$\begin{aligned} \psi_1{}^0 &= R_{01}(r)e^{-i\varphi} \\ \psi_2{}^0 &= R_{01}(r)e^{i\varphi} \end{aligned} \tag{6.16}$$

The radial part R_{01} of the zero order wave functions could be written out explicitly, but we shall not need it.

The elements of the perturbation matrix (6.15) include

$$b_{11} = \int (\psi_1{}^0)^* \mathscr{L}_z \psi_1{}^0 \, d\tau = \int (\psi_1{}^0)^* (\hbar/i) \frac{\partial}{\partial\varphi} \psi_1{}^0 \, d\tau = -\hbar$$

The student may wish to recover this result by explicit calculation using equations 4.7, but it follows directly from the fact that $\psi_1{}^0$ is an eigenfunction of the operator $\mathscr{L}_z$ with eigenvalue $\hbar$ (equation 4.10). Similarly the other elements of the perturbation matrix prove to be

$$b_{22} = \hbar; \qquad b_{12} = b_{21} = 0$$

so that the secular equation for the perturbed first excited state of the isotropic, two dimensional oscillator is

$$\begin{vmatrix} -\hbar - E^1 & 0 \\ 0 & \hbar - E^1 \end{vmatrix} = 0$$

This leads to first order perturbation energies and associated zero order wave functions

$$E_1 = E^0 - \varepsilon\hbar = 2h\nu - \varepsilon\hbar; \qquad \psi_1{}^0 = R_{01}e^{-i\varphi}$$

$$E_2 = E^0 + \varepsilon\hbar = 2h\nu + \varepsilon\hbar; \qquad \psi_2{}^0 = R_{01}e^{i\varphi}$$

In other words, the external magnetic field perturbs the oscillator in such a way as to split the first excited state into two different energy levels, each of which is associated to zero order with two now energetically distinguishable wave functions. The lower state E_1 corresponds to an alignment of the magnetic dipole of the oscillator with the field, and in the upper state E_2 its alignment is against the field. The reader may prefer to interpret this phenomenon in terms of the rotational motion of the oscillator: the external magnetic field favors, say, counterclockwise rotations; to perform clockwise rotations, the particle has to buck the field.

6.11 "BOND" FORMATION

Now let us consider the perturbation of the first excited state of the oscillator by a charged body brought to a distance D from the origin along the x axis (Figure 6.5). This is a fanciful model of the beginning of chemical bond formation. I suppose the perturbing potential of the foreign body to be a function only of its distance ρ from the moving oscillator. From the law of cosines

$$\rho^2 = r^2 + D^2 - 2rD\cos\varphi$$

so that the perturbing potential may be written $\varepsilon f(r, \cos\varphi)$ where ε is proportional to the charge on the foreign body. We shall not need to know the explicit form of f. It is sufficient for what we shall do here to recognize its

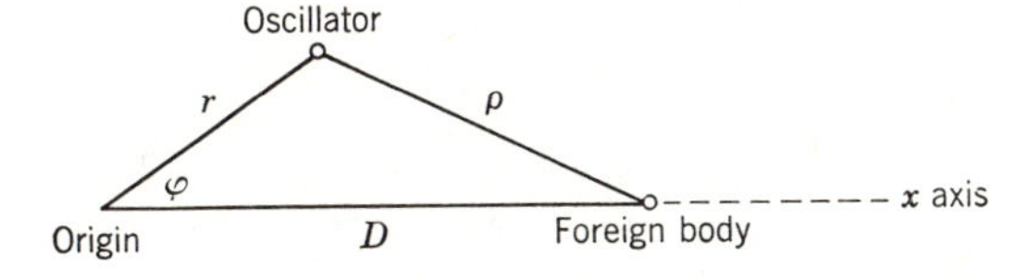

Figure 6.5 "Bond formation" in the two dimensional oscillator.

symmetry property, which consists of an indifference to a change of sign of φ. Physically this states that the point r, φ is equivalent in potential energy to the point r, $-\varphi$, or that the potential energy of the perturbed system is symmetric with respect to reflection through the x axis. Mathematically, f is an even function of φ.

Let us use as a basis for our perturbation calculation the same zero order wave functions (6.16) employed in Section 6.10. The required matrix elements are

$$\begin{aligned}
b_{11} &= \int (\psi_1{}^0)^* f(r, \cos\varphi)\psi_1{}^0 \, d\tau \\
&= \int_0^\infty rR_{01}{}^2 \, dr \int_0^{2\pi} e^{-i\varphi} f(r, \cos\varphi) e^{i\varphi} \, d\varphi \\
&= \int_0^\infty rR_{01}{}^2 \, dr \int_0^{2\pi} f(r, \cos\varphi) \, d\varphi \\
b_{22} &= b_{11} \\
b_{21} = b_{12} &= \int (\psi_1{}^0)^* f(r, \cos\varphi)\psi_2{}^0 \, d\tau \\
&= \int_0^\infty rR_{01}{}^2 \, dr \int_0^{2\pi} e^{-2i\varphi} f(r, \cos\varphi) \, d\varphi
\end{aligned} \tag{6.17}$$

We have as yet made no use of the symmetry property of f. In the last of the integrals (6.17) write $e^{-2i\varphi} = \cos 2\varphi - i \sin 2\varphi$ and note that $\cos 2\varphi$ is an even function of φ whereas $\sin 2\varphi$ is an odd function, changing sign when φ changes sign. Whatever the form of f, so long as f is an even function of φ the $\sin 2\varphi$ or imaginary part of the integrand vanishes upon integration over the full range 0 to 2π. We are left with the $\cos 2\varphi$ or real part of the integrand, so that the integral b_{12} is a real number. Let us call $b_{11} = b_{22} \equiv \alpha$ and $b_{12} = b_{21} \equiv \beta$. The perturbation matrix is

$$B = \begin{pmatrix} \alpha & \beta \\ \beta & \alpha \end{pmatrix}$$

with eigenvalues and eigenvectors

$$\begin{aligned}
\alpha + \beta; &\quad \left(\frac{1}{\sqrt{2}}, \frac{1}{\sqrt{2}}\right) \\
\alpha - \beta; &\quad \left(\frac{1}{\sqrt{2}}, -\frac{1}{\sqrt{2}}\right)
\end{aligned}$$

Now if $f(r, \cos\varphi)$ represents attraction, so that the oscillating particle is attracted to the foreign body, then f is everywhere negative and the integrals β and α must also be negative. It follows that the lower energy state is represented by the eigenvalue $\alpha + \beta$. The complete first order perturbation energies and their zero order wave functions are thus

$$E_1 = 2h\nu + \varepsilon(\alpha + \beta); \qquad \psi_x = \left(\frac{1}{\sqrt{2}}\right)(\psi_1{}^0 + \psi_2{}^0)$$

$$E_2 = 2h\nu + \varepsilon(\alpha - \beta); \qquad i\psi_y = \left(\frac{1}{\sqrt{2}}\right)(\psi_1{}^0 - \psi_2{}^0)$$

In other words, the attracting foreign body on the x axis has split the degenerate first excited state of the oscillator into two energy levels, the lower of which is associated with the real, zero order wave function ψ_x having a maximum concentration of charge along the x axis and the upper of which is associated with the real wave function ψ_y having a maximum concentration of charge along the y axis. (The factor i which multiplies ψ_y is actually superfluous in all calculations involving ψ_y and may be discarded.)

With regard to chemical bond formation it is extremely suggestive that this second calculation resolves the degeneracy of the unperturbed oscillator in such a way as to favor a zero order wave function with a *maximum concentration of charge directed towards the foreign body.*

PROBLEMS

6.2 Calculate the effect of a magnetic field (Zeeman effect) on the first excited state of the two dimensional, isotropic, harmonic oscillator, using as a basis the unperturbed functions (4.8):

$$\psi_x = R_{01}(r) \cos \varphi$$

$$\psi_y = R_{01}(r) \sin \varphi$$

Obtain to first order the new values of the split energy levels and assign to each level a zero order wave function which is a linear combination of ψ_x and ψ_y.

6.3

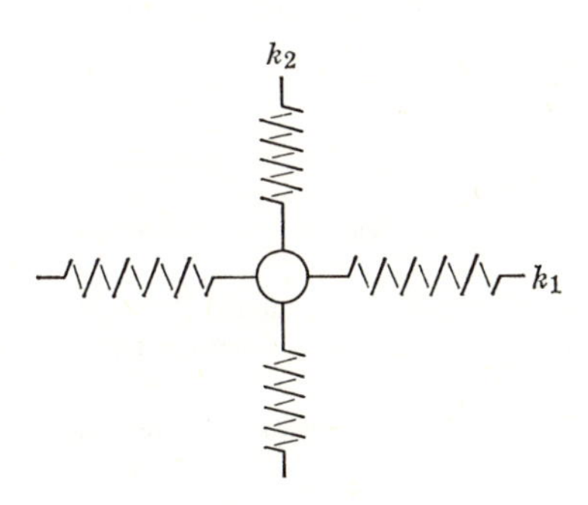

The two dimensional, anisotropic, harmonic oscillator is bound to the origin by two springs of different strength, k_1, k_2. Suppose that $k_1 = k$ and $k_2 = k + 2\varepsilon$ with ε small. Then the potential energy of the oscillator is

$$V = \tfrac{1}{2}k_1x^2 + \tfrac{1}{2}k_2y^2 = \tfrac{1}{2}k(x^2 + y^2) + \varepsilon y^2 = \tfrac{1}{2}kr^2 + \varepsilon r^2 \sin^2 \varphi$$

(1) Treat the second term in the potential energy as a perturbation and calculate to first order the perturbed energy levels derived from the first excited state of the isotropic oscillator. Use any suitable basis of your choice.

(2) To zero order of accuracy find appropriate wave functions for the perturbed levels.

(3) Investigate the commutation of the angular momentum operator with the Hamiltonian: does $\mathscr{L}_z H = H\mathscr{L}_z$? What is the quantum mechanical interpretation of this result?

6.4 The Zeeman effect for the hydrogen atom. The first excited state of the hydrogen atom is four fold degenerate: $E_2 = -m_e e^4/8\hbar^2$ and

$$2s = (4\sqrt{2\pi})^{-1}a_0^{-3/2}\left(2 - \frac{r}{a_0}\right)\exp\left(-\frac{r}{2a_0}\right)$$

$$2pz = (4\sqrt{2\pi})^{-1}a_0^{-3/2}\frac{r}{a_0}\exp\left(-\frac{r}{2a_0}\right)\cos\theta$$

$$2px = (4\sqrt{2\pi})^{-1}a_0^{-3/2}\frac{r}{a_0}\exp\left(-\frac{r}{2a_0}\right)\sin\theta\cos\varphi$$

$$2py = (4\sqrt{2\pi})^{-1}a_0^{-3/2}\frac{r}{a_0}\exp\left(-\frac{r}{2a_0}\right)\sin\theta\sin\varphi$$

in which a_0 is the Bohr radius. Calculate the new energy levels derived from the first excited state by a magnetic field $\mathscr{B}$ directed along the z axis. The perturbing operator is $H^1 = \varepsilon\mathscr{L}_z = \varepsilon(\hbar/i)(\partial/\partial\varphi)$ with ε proportional to the magnetic field strength, $\varepsilon = e\mathscr{B}/2m_ec$.

REFERENCES

1. H. Eyring, J. Walter, and G. E. Kimball, *Quantum Chemistry*, John Wiley and Sons, Inc., New York, 1944, p. 27.
2. W. Kauzmann, *Quantum Chemistry*, Academic Press, New York, 1957, p. 112.
3. A. Streitwieser, Jr., *Molecular Orbital Theory for Organic Chemists*, John Wiley and Sons, Inc., New York, 1961, Chapter 5.
4. A. L. McClellan, *Tables of Experimental Dipole Moments*, W. H. Freeman, San Francisco, 1963, p. 41.

Chapter 7

ELECTRON SPIN

7.1 HISTORY

The Bohr theory of the hydrogen atom and the Schrödinger theory that followed it started with a classical picture of the electron as a point charge. In view of the success of these theories in interpreting the spectrum of hydrogen, it was perfectly proper for the physicists of the time to imagine that the quantum mechanics of more complicated atoms could be managed in the same way, and early interpretations of atomic spectra were attempted on the basis of the electron as a point particle. Despite the success of much of this work, there were features of atomic spectra which could not be explained until the suggestion by Goudsmit and Uhlenbeck in 1925 that the electron in addition to its properties of mass and charge also acted like a tiny, permanent magnet. By that time physicists had learned to associate magnetism in atoms with the orbital motion of the electron around the nucleus visualized as an electric current, and they associated it in particular with the angular momentum operators after the manner discussed in Sections 4.14, 5.5 to 5.7, and 6.10. If the electron possesses an intrinsic magnetic moment independent of its orbital motion, then the classical picture of this being the result of current flowing in a circuit invoked the idea that the electron must be spinning around an internal axis after the manner of the earth.

7.2 THE RIGID ROTOR

A standard problem in many texts on quantum mechanics[1–3] is that of the rigid rotor. Let us for convenience imagine a solid, uniform sphere set into rotation about an internal axis. When interpreted quantum mechanically, this problem proves to be exactly solvable with energy eigenvalues

$$E_J = \frac{\hbar^2}{2I} J(J+1); \qquad J = 0, 1, 2, \ldots \tag{7.1}$$

in which I is a constant characteristic of the sphere known as its moment of inertia. The wave functions Y that are assigned to these levels are also eigenfunctions of the angular momentum operators

$$\begin{aligned} \mathscr{L}^2 Y_{Jm} &= J(J+1)\hbar^2 Y_{Jm} \\ \mathscr{L}_z Y_{Jm} &= m\hbar Y_{Jm}; \qquad m = -J, -J+1, \ldots, J \end{aligned} \tag{7.2}$$

The rigid rotor thus possesses an infinity of energy levels, each of which is $2J + 1$ fold degenerate.

These results have been successful in interpreting details in the infrared spectroscopy of rigid, rotating molecules, and one might reasonably expect that they should also apply to a spinning electron. That they do not was a source of mystification to the physicists of the time and suggests that the Schrödinger theory as we have so far developed it is incomplete. Indeed it was not until the work of Dirac in 1928 that the spin phenomenon was properly incorporated into quantum mechanics. Dirac showed that electron spin had no real classical analog in terms of a rotating, charged sphere. It was instead a requirement of the special theory of relativity, and the fact that the phenomenon nowadays is still referred to as "spin" is more for historical reasons than for the accuracy of the mental picture it invokes.

Fortunately it is possible to develop semiempirical rules for electron spin which, although they represent an artificial graft onto what we have up to this point developed logically, nevertheless work very well in chemical practice. It is only with a sense of apologetic inelegance that I must refer the reader to Dirac's theory[4,5] for the justification of these rules.

7.3 SPIN ONE HALF

Goudsmit and Uhlenbeck discovered empirically that small details in the spectra of the alkali atoms could be quantitatively explained by choosing for J in equation 7.1 the unique value $J = \frac{1}{2}$ and by permitting m to take on the two possibilities $m = -\frac{1}{2}, \frac{1}{2}$. Note that in the absence of an external field of force this restricts the freely spinning electron to a single rotational energy level, meaning that we shall never observe a spin transition to a different value of J. The angular momentum operators $\mathscr{L}^2$ and $\mathscr{L}_z$ for a spinning electron are conventionally written S^2 and S_z, and the unique value $J = \frac{1}{2}$ thus means that the total angular momentum $S = \sqrt{\frac{1}{2}(\frac{1}{2}+1)}\hbar$ is an invariant property of the electron. It follows that only the z component of angular momentum can shift between the possibilities $S_z = -\frac{1}{2}\hbar$ and $S_z = +\frac{1}{2}\hbar$. Together with its properties of charge and mass, the extra degree of freedom permitted to the electron by these two possible spin states is sufficient to explain all properties of the electron that are known up to the present time.

Despite the success of the spin one half hypothesis, however, it is nonetheless brutally true that $J = \frac{1}{2}$ is not an allowed quantum number for the rigid rotor and that the functions $Y_{\frac{1}{2}\frac{1}{2}}$, $Y_{\frac{1}{2}-\frac{1}{2}}$, which in the Schrödinger theory are supposed to be the corresponding wave functions, are not single valued over the complete range of orientational position coordinates of the rotor. The Schrödinger theory is thus incomplete, and the Dirac modifications were necessary to put the spin phenomenon into proper perspective.

7.4 ELECTRON SPIN RESONANCE SPECTROSCOPY

The fact that the rotational energy level of the electron is unique and invariant prohibits us from observing purely rotational transitions after the fashion possible for rotating molecules. If it were not for the doubly degenerate character of this single energy level, we should be unable to observe any spectra at all in connection with the isolated, spinning electron, and we can only observe transitions between these degenerate states if we make them slightly nondegenerate through the application of a magnetic field. The resulting Zeeman effect for the electron is analogous to that we worked out for the orbital motion of the isotropic oscillator in Section 6.10 and results in the splitting of the two spin states into two energy levels separated by an amount $2(e\mathscr{B}/m_e c)\hbar$ in which $-e$ and m_e are the charge and mass of the electron respectively, c the velocity of light, and $\mathscr{B}$ the strength of the magnetic field. This spin transition lies at the heart of electron spin resonance spectroscopy in which, say, a hydrogen atom is placed between the poles of a powerful magnet and subjected to electromagnetic radiation of radio frequency. By adjusting the strength of the magnetic field, a resonant transition between the two spin states of the electron can be observed, although from the point of view of the purely orbital motion of the electron, the state of the atom is always $1s$. The reason that this transition is of interest in chemistry lies in the fact that the electron "sees" not only the externally imposed magnetic field $\mathscr{B}$ but also fields of local, atomic or molecular origin. Thus the tiny magnetic fields created by the spinning nuclei of atoms and by the orbiting and spinning of other electrons in a molecule all superpose upon the $\mathscr{B}$ field produced by the apparatus with the result that electron spin resonance transitions occur at slightly different $\mathscr{B}$'s for different species, and this gives the chemist information about local magnetic environment in a molecular system.

7.5 A SPIN ALGEBRA

We shall now introduce an empirical spin algebra to be grafted on to the Schrödinger formalism. To avoid confusion with the magnetic quantum number m for the orbital motion of an electron in an atom, let me replace the m of Sections 7.2 and 7.3 by a spin quantum number s so that $s = \pm\frac{1}{2}$.

Despite the fact that there is no pair of Schrödinger type eigenfunctions of the rigid rotor corresponding to $J = \frac{1}{2}$; $s = \pm\frac{1}{2}$, I shall write down a formal algebra as though this were the case. Thus imagine that we have eigenfunctions α and β of the operator S_z so that formally

$$S_z\alpha = \tfrac{1}{2}\hbar\alpha$$

$$S_z\beta = -\tfrac{1}{2}\hbar\beta$$

The spin functions α and β are defined in a "spin space" whose coordinates I shall represent by a single letter ω. Thus because S_z is Hermitian, α and β are orthonormal, and

$$\begin{aligned} \int \alpha^*\alpha \, d\omega &= \int \beta^*\beta \, d\omega = 1 \\ \int \alpha^*\beta \, d\omega &= \int \beta^*\alpha \, d\omega = 0 \end{aligned} \tag{7.3}$$

We shall never require a more detailed knowledge of α and β than is expressed by these rules.

7.6 WAVE FUNCTIONS FOR HYDROGEN INCLUDING SPIN

If the electron is not isolated but is incorporated into a hydrogen atom, then its total Hamiltonian operator will be a function of the four coordinates r, θ, φ, ω. In the ordinary course of things we would expect this Hamiltonian to conform to the classical picture we have of satellite motion: there would be a term for the kinetic energy of orbital motion, another for the kinetic energy of spin motion, still another for the potential energy of electrostatic attraction of the electron for the nucleus. If the magnetic fields of the orbiting and of the spinning electron interact to an appreciable extent, there will also be a term representing this spin-orbit interaction. In actual fact, the second of these terms never appears, and no one has ever written an operator for the kinetic energy of a spinning electron. This again emphasizes the inadequacy of the classical picture invoked by the word "spin." The spin-orbit interaction does exist, however, and while we shall neglect it in this book, the reader should be aware that it gives rise to those anomalous features of the spectra of atoms whose detection led to the initial discovery of the spin phenomenon.

In the preceding paragraph we effectively knocked out of the Hamiltonian all those terms that involve the spin coordinate ω, and what remains is just the orbital part of the Hamiltonian which we have been using all along. To include the spin functions α and β in the final wave functions for hydrogen therefore involves no more than a little patch work. Because our atomic Hamiltonian $H(r, \theta, \varphi) = H(\tau)$ is free of the spin coordinate and the operator $S_z = S_z(\omega)$ involves only the spin coordinate, the two must commute

$$H(\tau)S_z(\omega) = S_z(\omega)H(\tau)$$

Hence if $\psi(\tau)$ is a solution of the orbital Schrödinger equation $H(\tau)\psi(\tau) = E\psi(\tau)$, wave functions that are simultaneously eigenfunctions of both H and S_z are

$$\psi(\tau, \tfrac{1}{2}) = \psi(\tau)\alpha(\omega)$$

$$\psi(\tau, -\tfrac{1}{2}) = \psi(\tau)\beta(\omega)$$

These functions are degenerate and orthogonal, for

$$H(\tau)\psi(\tau, \tfrac{1}{2}) = H(\tau)\psi(\tau)\alpha(\omega) = \alpha(\omega)H(\tau)\psi(\tau) = E\psi(\tau)\alpha(\omega) = E\psi(\tau, \tfrac{1}{2})$$

$$H(\tau)\psi(\tau, -\tfrac{1}{2}) = H(\tau)\psi(\tau)\beta(\omega) = \beta(\omega)H(\tau)\psi(\tau) = E\psi(\tau)\beta(\omega) = E\psi(\tau, -\tfrac{1}{2})$$

and from our rules (7.3) we have

$$\iint \psi^*(\tau, \tfrac{1}{2})\psi(\tau, -\tfrac{1}{2})\, d\tau\, d\omega = \iint \psi^*(\tau)\alpha^*(\omega)\psi(\tau)\beta(\omega)\, d\tau\, d\omega$$

$$= \int \psi^*(\tau)\psi(\tau)\, d\tau \int \alpha^*\beta\, d\omega = 1 \times 0 = 0$$

Complete wave functions for hydrogen are this very easy to construct. We have only to take the $1s$, $2s$, $2p$, . . . orbital functions derived in Chapter 5 and multiply them by either α or β to obtain wave functions including the spin. No energy levels will be changed in value, and the only apparent effect of all of this is that the total number of wave functions is doubled. For example, the general hydrogen wave function $\psi(n, l, m, s)$ yields two degenerate choices for the ground state of hydrogen

$$\psi(1, 0, 0, \tfrac{1}{2}) = \psi_{1s}(r, \theta, \varphi)\alpha(\omega)$$

$$\psi(1, 0, 0, -\tfrac{1}{2}) = \psi_{1s}(r, \theta, \varphi)\beta(\omega)$$

The first is referred to colloquially as "spin up" ($s = \frac{1}{2}$, pictorially $\uparrow$) and the second as "spin down" ($s = -\frac{1}{2}$, pictorially $\downarrow$). Both wave functions possess an orbital energy $E_{1s} = -313.8$ kcal/mole. In a similar way with the inclusion of spin the degeneracy of the first excited state of the hydrogen atom increases from four to eight, for each of the four orbitally degenerate functions $2s$, $2p_{-1}$, $2p_0$, $2p_1$ may be multiplied by either α or β.

7.7 FINAL APOLOGY

The perceptive reader cannot but be left unsatisfied by patchwork of this sort, for it does not flow in a logically acceptable way from the postulates of Section 4.6. I would encourage to the student seriously interested in quantum mechanics to nourish his dissatisfaction to the extent that he will make a deeper investigation into the place of spin in quantum mechanics by reading what the physicists have to say about it.[4,5] So far as chemistry is concerned, however, the rules laid down here for the construction of spin dependent wave functions work very well in practice, and that must be the ultimate test of the usefulness of any theory.

PROBLEM

7.1 The angular momentum of an electron in a hydrogen atom is of order $\hbar$ and its distance from the nucleus is of the order a_0 = one Bohr radius. Imagining the electron to be moving in a circular track, its classical speed v would then be of the order $v = \hbar/a_0 m_e$ in which m_e is the mass of the electron. Relativistic effects exclusive of spin increase with the ratio v^2/c^2 in which c is the velocity of light. Estimate the ratio v^2/c^2 for an electron in a hydrogen atom and determine the suitability of the nonrelativistic Schrödinger equation for hydrogen.

REFERENCES

1. W. Kauzmann, *Quantum Chemistry*, Academic Press, New York, 1957, pp. 198–201.
2. H. Eyring, J. Walter, and G. E. Kimball, *Quantum Chemistry*, John Wiley and Sons, Inc., New York, 1944, pp. 72–75.
3. F. L. Pilar, *Elementary Quantum Chemistry*, McGraw-Hill, Inc., New York, 1968, pp. 168–171.
4. P. A. M. Dirac, *The Principles of Quantum Mechanics*, Oxford Press, 1958, Chapter 11.
5. H. A. Kramers, *Quantum Mechanics*, North-Holland Press, Amsterdam, 1957, Chapter 6.

Chapter 8

ATOMIC STRUCTURE

8.1 INTRODUCTION

It has proved impossible to solve the Schrödinger equation analytically for any electrically neutral atomic system more complicated than hydrogen. Since this leaves over 99% of the periodic table unaccounted for, the reader can easily appreciate the importance of approximate methods in classifying quantum mechanical results for atoms, and the perturbation theory developed in Chapter 6 will assume a position of paramount importance in this one. While not exact, the approximate results obtained in this way have proved to be uniquely valuable in correlating the observed experimental facts. Perturbation theory by itself, however, has proved inadequate to the task of explaining all the observations, even approximately. A new, fundamental law of nature must, therefore, be incorporated into quantum mechanics when it is applied to systems of many electrons: the Pauli exclusion principle.

8.2 THE HELIUM ATOM

We can rapidly describe the features of the quantum mechanical system consisting of two electrons moving together in a common central field of charge $+2e$. Because we can always incorporate spin into the picture after having first solved the orbital problem, we shall at the start go on treating the electron as a point charge. The configuration space will thus be $2 \times 3 = 6$ dimensional, and when both electrons are referred to a set of spherical polar coordinates centered on the nucleus, these coordinates are $r_1, \theta_1, \varphi_1; r_2, \theta_2, \varphi_2$, the subscript indicating a labeling of the electrons as 1 and 2. The classical Hamiltonian function exclusive of spin is the sum of the kinetic energies of both electrons plus the potential energy of each resulting from their attraction

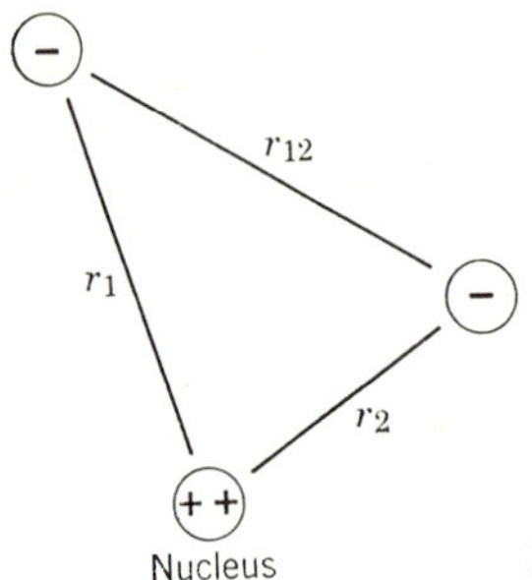

Figure 8.1 The helium atom.

for the atomic nucleus plus a potential energy term resulting from their mutual electrostatic repulsion.

When these features of the classical Hamiltonian are translated through the quantum rules into a Hamiltonian operator, we have, referring to Figure 8.1,

$$H = -\frac{\hbar^2}{2m_e}(\nabla_1^2 + \nabla_2^2) - \frac{2e^2}{r_1} - \frac{2e^2}{r_2} + \frac{e^2}{r_{12}} \tag{8.1}$$

The kinetic energy part contains a Laplacian operator ∇^2 expressed in the coordinates of each electron, and the Coulomb repulsion term e^2/r_{12} must for practical calculations be rewritten as a complicated expansion in our six basic coordinates. It is the presence of the e^2/r_{12} term that prevents separation of the Schrödinger equation for helium and renders an analytic solution inaccessible.

8.3 THE GROUND STATE OF THE HELIUM ATOM

Because we despair of finding an analytic solution to the six dimensional partial differential equation implied by (8.1), we search for approximate results. These may be obtained from a perturbation calculation if we solve exactly the simpler equation obtained by dropping from the Hamiltonian the complicated quantity e^2/r_{12} and then bring it back in again as a perturbation. Thus we set $H = H^0 + \varepsilon H^1$ with

$$H^0 = -\frac{\hbar^2}{2m_e}(\nabla_1^2 + \nabla_2^2) - \frac{2e^2}{r_1} - \frac{2e^2}{r_2}$$

$$H^1 = \frac{1}{r_{12}}$$

$$\varepsilon = e^2$$

This zero order approximation we shall call the independent electron model, for it neglects all interaction between the electrons.

Using the simplified Hamiltonian H^0, the Schrödinger equation separates immediately into two partial differential equations in three dimensions each if we assume that the zero order wave function may be factored

$$\psi^0(1, 2) = \psi'(1)\psi''(2)$$

into two functions, each of which involves the coordinates of a single electron only. Let

$$H^0 = H(1) + H(2)$$

where

$$H(1) = -\frac{\hbar^2}{2m_e}\nabla_1^{\,2} - \frac{2e^2}{r_1}$$

is the Hamiltonian operator for a single electron moving in a central field of charge $2e$, and similarly for $H(2)$. The reader should satisfy himself that the separated equations are

$$H(1)\psi'(1) = E'\psi'(1)$$
$$H(2)\psi''(2) = E''\psi''(2)$$

with $E^0 = E' + E''$. Each of these equations takes "hydrogen-like" solutions, the only difference being the increase in nuclear charge. For the ground state, the assignment of orbital quantum numbers for both electrons is $n = 1$, $l = 0$, $m = 0$ with an attendant single electron energy of $E' = E'' = -2m_e e^4/\hbar^2 = -1255$ kcal. For the helium atom in the ground state we have therefore a zero order energy $E^0 = -2(1255) = -2510$ kcal and an orbitally nondegenerate zero order wave function

$$\psi^0(1, 2) = \psi_{1s}(1)\psi_{1s}(2) \tag{8.2}$$

where

$$\psi_{1s} = \frac{1}{\sqrt{\pi}}\left(\frac{2}{a_0}\right)^{3/2} \exp\left(-\frac{2r}{a_0}\right)$$

In the absence of any knowledge whatsoever of electron spin, we would proceed from this point to calculate an approximate ground state energy for the helium atom by using (8.2) in a nondegenerate perturbation integral J,

$$E = E^0 + e^2J$$

with

$$J = \int (\psi^0)^* H^1 \psi^0 \, d\tau$$

$$= \iint \psi_{1s}^*(1)\psi_{1s}^*(2) \frac{1}{r_{12}} \psi_{1s}(1)\psi_{1s}(2) \, d\tau_1 \, d\tau_2$$

$$= \iint [\psi_{1s}(1)\psi_{1s}(2)]^2 \frac{1}{r_{12}} \, d\tau_1 \, d\tau_2$$

$$= \int_0^\infty r_1^2 \, dr_1 \int_0^\pi \sin\theta_1 \, d\theta_1 \int_0^{2\pi} d\varphi_1 \int_0^\infty r_2^2 \, dr_2 \int_0^\pi \sin\theta_2 \, d\theta_2$$

$$\times \int_0^{2\pi} d\varphi_2 \left(\frac{1}{\pi}\right)^2 \left(\frac{2}{a_0}\right)^6 \exp\left[-\frac{4}{a_0}(r_1 + r_2)\right] \frac{1}{r_{12}}$$

Into this last sixfold integral must be inserted the previously mentioned complicated expansion for $1/r_{12}$ before the integrations can be performed*; but courage fails me, and I shall not inflict this bit of mathematical horror upon the student. The kind gentleman† who first worked this out, however, reports that $J = \frac{5}{4}a_0$, so that to first order the energy of the ground state of the helium atom is

$$E = E^0 + \frac{5e^2}{4a_0} = -1726 \text{ kcal}$$

Experimentally it is found to be $E = -1822$ kcal.

8.4 THE GROUND STATE OF HELIUM INCLUDING SPIN

So far we would seem to be on reasonably firm ground, with the results of a simple perturbation calculation not far off the mark from the experimentally observed ground state energy. Some disturbing features begin to creep in, however, if we patch up our wave functions to take account of electron spin. Let us return to the single electron functions $\psi_{1s}(1)$ and $\psi_{1s}(2)$ and append to them the factors α and β which we found in Section 7.6 to be necessary for the spin dependent, single electron functions of hydrogen:

$$\text{electron 1:} \quad \{\psi_{1s}(1)\alpha(1); \qquad \psi_{1s}(1)\beta(1)\}$$
$$\text{electron 2:} \quad \{\psi_{1s}(2)\alpha(2); \qquad \psi_{1s}(2)\beta(2)\}$$

* The integral can also be evaluated by an electrostatic argument. See reference 5, page 446.

† A. Unsöld, *Ann. Physik*, **82,** 355 (1927).

A product of any member of the first row times any member of the second row is a solution of the independent electron Schrödinger equation for helium, and because there are four such products,

$$\begin{aligned} \psi_{\mathrm{I}}^0 &= \psi_{1s}(1)\psi_{1s}(2)\alpha(1)\alpha(2) && \uparrow\uparrow \\ \psi_{\mathrm{II}}^0 &= \psi_{1s}(1)\psi_{1s}(2)\alpha(1)\beta(2) && \uparrow\downarrow \\ \psi_{\mathrm{III}}^0 &= \psi_{1s}(1)\psi_{1s}(2)\beta(1)\alpha(2) && \downarrow\uparrow \\ \psi_{\mathrm{IV}}^0 &= \psi_{1s}(1)\psi_{1s}(2)\beta(1)\beta(2) && \downarrow\downarrow \end{aligned} \tag{8.3}$$

it would seem that with the inclusion of spin the ground state energy of helium should be estimated using degenerate perturbation theory rather than the nondegenerate perturbation calculation of Section 8.3. To do this we shall require the elements of the 4 × 4 perturbation matrix

$$H_{ij} = \iint (\psi_i^0)^* \frac{1}{r_{12}} \psi_j^0 \, d\tau \, d\omega$$

and I give H_{11} and H_{12} as examples:

$$\begin{aligned} H_{11} &= \iint (\psi_{\mathrm{I}}^0)^* \frac{1}{r_{12}} \psi_{\mathrm{I}}^0 \, d\tau \, d\omega \\ &= \int [\psi_{1s}(1)\psi_{1s}(2)]^2 \frac{1}{r_{12}} \, d\tau \int \alpha^*(1)\alpha^*(2)\alpha(1)\alpha(2) \, d\omega \\ &= \int [\psi_{1s}(1)\psi_{1s}(2)]^2 \frac{1}{r_{12}} \, d\tau \int \alpha^*(1)\alpha(1) \, d\omega_1 \int \alpha^*(2)\alpha(2) \, d\omega_2 \\ &= \left\{ \int [\psi_{1s}(1)\psi_{1s}(2)]^2 \frac{1}{r_{12}} \, d\tau \right\} \times 1 \times 1 = J \\ H_{12} &= \iint (\psi_{\mathrm{I}}^0)^* \frac{1}{r_{12}} \psi_{\mathrm{II}}^0 \, d\tau \, d\omega \\ &= \int [\psi_{1s}(1)\psi_{1s}(2)]^2 \frac{1}{r_{12}} \, d\tau \int \alpha^*(1)\alpha^*(2)\alpha(1)\beta(2) \, d\omega \\ &= \int [\psi_{1s}(1)\psi_{1s}(2)]^2 \frac{1}{r_{12}} \, d\tau \int \alpha^*(1)\alpha(1) \, d\omega_1 \int \alpha^*(2)\beta(2) \, d\omega_2 \\ &= J \times 1 \times 0 = 0 \end{aligned}$$

The matrix element H_{12} therefore vanishes because the functions ψ_{I}^0 and ψ_{II}^0 are "spin orthogonal." In a similar way the reader may confirm for himself the fact that the perturbation matrix has elements $H_{ii} = J$ and $H_{ij} = 0$ for

$i \neq j$. The perturbed energy levels must then be calculated from the eigenvalues of the matrix

$$\begin{pmatrix} J & 0 & 0 & 0 \\ 0 & J & 0 & 0 \\ 0 & 0 & J & 0 \\ 0 & 0 & 0 & J \end{pmatrix}$$

which is to say that the degeneracy is not removed by interelectronic repulsion, and all of the four zero order wave functions for the ground state of the helium atom are perturbed to the same extent. So far this is in agreement with what we have found before, and the ground state of the helium atom from this calculation has a first order perturbation energy $E = E^0 + e^2J = E^0 + 5e^2/4a_0$, but also according to this the ground state is fourfold degenerate, whereas by neglecting spin our previous study predicted it to be nondegenerate. It is enough to make any conscientious theorist nervous when all experimental evidence points to the fact that the ground state of helium is actually nondegenerate.

8.5 SYMMETRIC AND ANTISYMMETRIC WAVE FUNCTIONS

The theorems of linear algebra supported by our experience with Problems 6.1 and 6.2 indicate that in performing a degenerate perturbation calculation, the choice of a basis of zero order degenerate vectors or zero order degenerate wave functions is a matter of convenience. The results of such perturbation calculations are always the same no matter what our initial basis, and only the computational details are different. We shall not therefore change anything in the physics of the situation if instead of the basis 8.3 for our ground state helium calculation we employ a new basis synthesized out of the old by making orthogonal linear combinations of the functions (8.3). The motivation for doing this must remain obscure for the time being, but to lay a foundation for the Pauli principle to come, I will state that the new functions will have the property that they are eigenfunctions of the permutation operator which changes the numbering of the electrons. A permutation operator of this sort has no analog in anything we have come across so far, largely for the fact that it is only in the helium atom that we have had for the first time to deal with a system comprising more than one electron. On intuitive grounds, however, it would seem not unreasonable to demand that if in our mathematics we interchange the coordinates of electrons 1 and 2, the resulting atom will be physically indistinguishable from the old.

Turn back to the functions (8.3) and interchange electron coordinates.

Denoting this interchange by a permutation symbol P we have formally

$$P\psi_j{}^0(1, 2) = \psi_j{}^0(2, 1)$$

or in detail

$$P\psi_{\rm I}^0 = P[\psi_{1s}(1)\psi_{1s}(2)\alpha(1)\alpha(2)] = \psi_{1s}(2)\psi_{1s}(1)\alpha(2)\alpha(1) = \psi_{\rm I}^0$$

The interchange of coordinates in $\psi_{\rm I}^0$ is thus seen to have no effect, and we can state this formally by remarking that $\psi_{\rm I}^0$ is an eigenfunction of P with eigenvalue 1.

For $\psi_{\rm II}^0$ the situation is not so simple:

$$P\psi_{\rm II}^0 = P[\psi_{1s}(1)\psi_{1s}(2)\alpha(1)\beta(2)] = \psi_{1s}(2)\psi_{1s}(1)\alpha(2)\beta(1) \equiv \psi_{\rm III}^0$$

or in brief

$$P\psi_{\rm II}^0 = \psi_{\rm III}^0$$

so that $\psi_{\rm II}^0$ is not an eigenfunction of P. Continuing in this way the student will readily confirm for himself the operations

$$P\psi_{\rm III}^0 = \psi_{\rm II}^0$$
$$P\psi_{\rm IV}^0 = \psi_{\rm IV}^0$$

We perceive that of our basis set (8.3), only $\psi_{\rm I}^0$ and $\psi_{\rm IV}^0$ are eigenfunctions of P, each with eigenvalue 1. To transform our basis set into a new one whose members are eigenfunctions of P, we have therefore only to tamper with $\psi_{\rm II}^0$ and $\psi_{\rm III}^0$ by setting up orthogonal linear combinations:

$$\begin{aligned}
\psi_2{}^0 &= \frac{1}{\sqrt{2}}(\psi_{\rm II}^0 + \psi_{\rm III}^0) = \psi_{1s}(1)\psi_{1s}(2)\frac{1}{\sqrt{2}}[\alpha(1)\beta(2) + \alpha(2)\beta(1)] \\
\psi_3{}^0 &= \frac{1}{\sqrt{2}}(\psi_{\rm II}^0 - \psi_{\rm III}^0) = \psi_{1s}(1)\psi_{1s}(2)\frac{1}{\sqrt{2}}[\alpha(1)\beta(2) - \alpha(2)\beta(1)]
\end{aligned} \tag{8.4}$$

By direct experimentation the student will confirm the operational equations

$$P\psi_2{}^0 = \psi_2{}^0$$

$$P\psi_3{}^0 = -\psi_3{}^0$$

so that of the four functions

$$\psi_1{}^0 = \psi_{\rm I}^0 = \psi_{1s}(1)\psi_{1s}(2)\alpha(1)\alpha(2)$$

$$\psi_2{}^0 = \psi_{1s}(1)\psi_{1s}(2)\frac{1}{\sqrt{2}}[\alpha(1)\beta(2) + \alpha(2)\beta(1)]$$

$$\psi_3{}^0 = \psi_{1s}(1)\psi_{1s}(2)\frac{1}{\sqrt{2}}[\alpha(1)\beta(2) - \alpha(2)\beta(1)]$$

$$\psi_4{}^0 = \psi_{\rm IV}^0 = \psi_{1s}(1)\psi_{1s}(2)\beta(1)\beta(2)$$

the first two and the last are eigenfunctions of P with eigenvalue 1, while the third is an eigenfunction of P with eigenvalue -1. An obvious classification scheme has led to the names symmetric and antisymmetric functions for these sets. As usual the student should note that the transformation from one basis set to another has been performed with an orthogonal matrix in which the four zero order wave functions transform like the components of a vector:

$$\begin{pmatrix} \psi_1{}^0 \\ \psi_2{}^0 \\ \psi_3{}^0 \\ \psi_4{}^0 \end{pmatrix} = \begin{pmatrix} 1 & 0 & 0 & 0 \\ 0 & \dfrac{1}{\sqrt{2}} & \dfrac{1}{\sqrt{2}} & 0 \\ 0 & \dfrac{1}{\sqrt{2}} & -\dfrac{1}{\sqrt{2}} & 0 \\ 0 & 0 & 0 & 1 \end{pmatrix} \begin{pmatrix} \psi_{\rm I}^0 \\ \psi_{\rm II}^0 \\ \psi_{\rm III}^0 \\ \psi_{\rm IV}^0 \end{pmatrix}$$

We have hardly a sufficient theoretical basis for taking the next step, but it is suggestive that the antisymmetric set in this classification has only one member, so that if somehow nature prohibited all members of the symmetric set, then the ground state of helium would indeed be nondegenerate with a zero order (independent electron) wave function

$$\psi^0(1, 2) = \psi_3{}^0 = \psi_{1s}(1)\psi_{1s}(2)\frac{1}{\sqrt{2}}[\alpha(1)\beta(2) - \alpha(2)\beta(1)]$$

This turns out to be precisely the procedure demanded by the Pauli principle. For future reference, let us note that the antisymmetric independent electron wave function may be written as a determinant

$$\psi_3{}^0 = \frac{1}{\sqrt{2}}\begin{vmatrix} \psi_{1s}(1)\alpha(1) & \psi_{1s}(2)\alpha(2) \\ \psi_{1s}(1)\beta(1) & \psi_{1s}(2)\beta(2) \end{vmatrix} \tag{8.5}$$

8.6 THE FIRST EXCITED STATE OF HELIUM

Let us push our calculations to a study of the first excited state of the helium atom. The details will be simplified (and more accurate) if we choose our zero order wave functions from a slightly modified scheme: we still want to divide the Hamiltonian into two parts $H = H^0 + \varepsilon H^1$ in which the part H^0 leads to an exactly solvable problem and εH^1 is small. Up to now we have done this by letting H^0 be the sum of those parts of the complete Hamiltonian due to the kinetic energy of the electrons plus the potential energy of electron-nucleus attraction. This led to a hydrogen-like zero order Hamiltonian. The results can be improved if we imagine each electron to be moving not in the bare field of the nucleus but in a radially symmetric field due to the nucleus plus the smeared out electron density of the other electron. The resulting zero order problem is still a central field one, but the field is no longer that of pure Coulomb attraction. Specifically this means that we write the potential energy for each electron

$$V = V^0 + V^1$$
$$= \left[-\frac{2e^2}{r} + W(r)\right] + \left[\frac{e^2}{r_{12}} - W(r)\right]$$

and use the first bracket for our zero order calculation and the second for the perturbation term. If we are skilful in the choice of $W(r)$, the results of first order perturbation theory should be improved. This thought is the inspiration of the Hartree and Hartree-Fock calculations on many electron atoms.[2–4]

We need in this text never be concerned with the specific choice of $W(r)$, for none of our calculations will be so detailed as to compel a choice. The importance of the method to our work, however, lies in the fact that once the central field problem is solved for the non-Coulomb potential $V^0 = -2e^2/r + W(r)$, only the radial factor of the single electron wave function is affected. The angular part is unaltered and separates off as before, and the angular momentum operators still commute with the zero order Hamiltonian. Most importantly, *the energy levels are found to depend on the quantum number l as well as on the principal quantum number n.* We thus remove much of the degeneracy involved in the use of hydrogen-like wave functions for perturbation calculations and also improve the accuracy of our zero order wave functions.

As an example of the use of this trick, as long as hydrogen-like wave functions are used, the first excited state for a helium atom would be constructed from zero order orbital wave functions corresponding to one electron in the $1s$ level and the second electron in any of the $2s$, $2p_{-1}$, $2p_0$, $2p_1$ levels. By introducing a properly chosen non-Coulomb central field as the zero order problem, the $2s$ level turns out to have a different (and lower) energy from any of the $2p$ levels, so that our problem is simplified to a perturbation calculation involving only those zero order wave functions which can be constructed from the presence of one electron in a $1s$ level and another in a $2s$ level. The short hand notation used for this configuration is $(1s)(2s)$.

Zero order wave functions for the configuration $(1s)(2s)$ are

$$\psi_{\text{I}}^0 = \psi_{1s}(1)\psi_{2s}(2)\alpha(1)\alpha(2) \qquad \psi_{\text{II}}^0 = \psi_{1s}(2)\psi_{2s}(1)\alpha(2)\alpha(1)$$
$$\psi_{\text{III}}^0 = \psi_{1s}(1)\psi_{2s}(2)\alpha(1)\beta(2) \qquad \psi_{\text{IV}}^0 = \psi_{1s}(2)\psi_{2s}(1)\alpha(2)\beta(1)$$
$$\psi_{\text{V}}^0 = \psi_{1s}(1)\psi_{2s}(2)\beta(1)\alpha(2) \qquad \psi_{\text{VI}}^0 = \psi_{1s}(2)\psi_{2s}(1)\beta(2)\alpha(1)$$
$$\psi_{\text{VII}}^0 = \psi_{1s}(1)\psi_{2s}(2)\beta(1)\beta(2) \qquad \psi_{\text{VIII}}^0 = \psi_{1s}(2)\psi_{2s}(1)\beta(2)\beta(1)$$

The four functions in the left hand column are analogous to the four (8.3) we derived for the ground state of helium. They are obtained by making all possible spin assignments to electron 1 in the $1s$ level and electron 2 in the $2s$ level. The second column of four functions is identical with the first except that electrons 1 and 2 have interchanged their roles:

$$P\psi_{\mathrm{I}}^0 = \psi_{\mathrm{II}}^0$$
$$P\psi_{\mathrm{III}}^0 = \psi_{\mathrm{IV}}^0$$
$$P\psi_{\mathrm{V}}^0 = \psi_{\mathrm{VI}}^0$$
$$P\psi_{\mathrm{VII}}^0 = \psi_{\mathrm{VIII}}^0$$

For the ground state of helium such a permutation of coordinates did not lead to an increase in the total degeneracy of the ground state or $(1s)^2$ configuration, but here it gives us four new functions for a grand total of eight. This amplification in the number of zero order wave functions due to the permutation of electron coordinates we may call "exchange degeneracy."

Now let us carry out a change of basis which will make all zero order functions into eigenfunctions of P. Remember that as long as the new basis is orthonormal, it will be just as serviceable a starting point for a perturbation calculation as any other degenerate set of eigenfunctions of the zero order, independent electron Hamiltonian H^0. I construct

$$\psi_1{}^0 = \frac{1}{\sqrt{2}}(\psi_{\mathrm{I}}^0 - \psi_{\mathrm{II}}^0) = \frac{1}{\sqrt{2}}[\psi_{1s}(1)\psi_{2s}(2) - \psi_{1s}(2)\psi_{2s}(1)]\alpha(1)\alpha(2)$$

$$\psi_2{}^0 = \frac{1}{\sqrt{2}}(\psi_{\mathrm{I}}^0 + \psi_{\mathrm{II}}^0) = \frac{1}{\sqrt{2}}[\psi_{1s}(1)\psi_{2s}(2) + \psi_{1s}(2)\psi_{2s}(1)]\alpha(1)\alpha(2)$$

$$\psi_3{}^0 = \frac{1}{\sqrt{2}}(\psi_{\mathrm{III}}^0 - \psi_{\mathrm{IV}}^0)$$

$$\psi_4{}^0 = \frac{1}{\sqrt{2}}(\psi_{\mathrm{III}}^0 + \psi_{\mathrm{IV}}^0)$$

$$\psi_5{}^0 = \frac{1}{\sqrt{2}}(\psi_{\mathrm{V}}^0 - \psi_{\mathrm{VI}}^0)$$

$$\psi_6{}^0 = \frac{1}{\sqrt{2}}(\psi_{\mathrm{V}}^0 + \psi_{\mathrm{VI}}^0)$$

$$\psi_7{}^0 = \frac{1}{\sqrt{2}}(\psi_{\mathrm{VII}}^0 - \psi_{\mathrm{VIII}}^0)$$

$$\psi_8{}^0 = \frac{1}{\sqrt{2}}(\psi_{\mathrm{VII}}^0 + \psi_{\mathrm{VIII}}^0)$$

Written in matrix language, this change of basis may be recognized as a linear transformation with an orthogonal matrix in an eight dimensional vector space

$$\begin{pmatrix} \psi_1^0 \\ \psi_2^0 \\ \psi_3^0 \\ \psi_4^0 \\ \psi_5^0 \\ \psi_6^0 \\ \psi_7^0 \\ \psi_8^0 \end{pmatrix} = \begin{pmatrix} 1/\sqrt{2} & -1/\sqrt{2} & & & & & & \\ 1/\sqrt{2} & 1/\sqrt{2} & & & & & & \\ & & 1/\sqrt{2} & -1/\sqrt{2} & & & & \\ & & 1/\sqrt{2} & 1/\sqrt{2} & & & & \\ & & & & 1/\sqrt{2} & -1/\sqrt{2} & & \\ & & & & 1/\sqrt{2} & 1/\sqrt{2} & & \\ & & & & & & 1/\sqrt{2} & -1/\sqrt{2} \\ & & & & & & 1/\sqrt{2} & 1/\sqrt{2} \end{pmatrix} \begin{pmatrix} \psi_{\mathrm{I}}^0 \\ \psi_{\mathrm{II}}^0 \\ \psi_{\mathrm{III}}^0 \\ \psi_{\mathrm{IV}}^0 \\ \psi_{\mathrm{V}}^0 \\ \psi_{\mathrm{VI}}^0 \\ \psi_{\mathrm{VII}}^0 \\ \psi_{\mathrm{VIII}}^0 \end{pmatrix}$$

In addition to being eigenfunctions of H^0, the student will confirm that the new basis functions are simultaneously eigenfunctions of P,

$$\begin{aligned} P\psi_1^0 &= -\psi_1^0 & \quad P\psi_2^0 &= \psi_2^0 \\ P\psi_3^0 &= -\psi_3^0 & \quad P\psi_4^0 &= \psi_4^0 \\ P\psi_5^0 &= -\psi_5^0 & \quad P\psi_6^0 &= \psi_6^0 \\ P\psi_7^0 &= -\psi_7^0 & \quad P\psi_8^0 &= \psi_8^0 \end{aligned}$$

those with odd index being the antisymmetric set and those with even index the symmetric set. Note also that every member of the antisymmetric set can be written as a 2×2 determinant of the type (8.5), such as

$$\psi_1^0 = \frac{1}{\sqrt{2}} \begin{vmatrix} \psi_{1s}(1)\alpha(1) & \psi_{1s}(2)\alpha(2) \\ \psi_{2s}(1)\alpha(1) & \psi_{2s}(2)\alpha(2) \end{vmatrix}$$

Having now constructed our basis for a first order perturbation calculation on a degenerate level, we set up the matrix with elements

$$H_{ij} = \int (\psi_i^0)^* f(1, 2) \psi_j^0 \, d\tau \, d\omega$$

where $f(1, 2) = 1/r_{12} - W(r_1) - W(r_2)$, and find its eigenvalues. The matrix will be 8×8 in size, but because $H_{ij} = 0$ if ψ_i^0 and ψ_j^0 belong to sets of different symmetry, the secular equation factors into two determinants of order 4 each. As an example of this phenomenon consider

$$
\begin{aligned}
H_{12} &= \int (\psi_1{}^0)^* f(1,2) \psi_2{}^0 \, d\tau \, d\omega \\
&= \tfrac{1}{2} \int [\psi_{1s}(1)\psi_{2s}(2) - \psi_{1s}(2)\psi_{2s}(1)]^* f(1,2) [\psi_{1s}(1)\psi_{2s}(2) + \psi_{1s}(2)\psi_{2s}(1)] \, d\tau \\
&\quad \times \int \alpha^*(1)\alpha^*(2)\alpha(1)\alpha(2) \, d\omega \\
&= \tfrac{1}{2} \int [\psi_{1s}(1)\psi_{2s}(2)]^2 f(1,2) \, d\tau - \tfrac{1}{2} \int [\psi_{1s}(2)\psi_{2s}(1)]^2 f(1,2) \, d\tau \\
&\quad + \tfrac{1}{2} \int \psi_{1s}(1)\psi_{2s}(2) f(1,2) \psi_{1s}(2)\psi_{2s}(1) \, d\tau \\
&\quad - \tfrac{1}{2} \int \psi_{1s}(2)\psi_{2s}(1) f(1,2) \psi_{1s}(1)\psi_{2s}(2) \, d\tau
\end{aligned}
$$

Because $f(1, 2)$ is symmetric with respect to interchanges of electron coordinates, the first and second integrals cancel, as do the third and fourth. Explicitly for

$$
\begin{aligned}
J &= \int [\psi_{1s}(1)\psi_{2s}(2)]^2 f(1,2) \, d\tau \qquad \text{(called a Coulomb integral)} \\
K &= \int \psi_{1s}(1)\psi_{2s}(2) f(1,2) \psi_{1s}(2)\psi_{2s}(1) \, d\tau \qquad \text{(called an exchange integral)}
\end{aligned}
\tag{8.6}
$$

the matrix element H_{12} is

$$H_{12} = \tfrac{1}{2}J - \tfrac{1}{2}J + \tfrac{1}{2}K - \tfrac{1}{2}K = 0$$

The same result will be found to apply to all H_{ij} formed from wave functions $\psi_i{}^0$ and $\psi_j{}^0$ of different symmetry. The appearance of our secular equation is thus that of an 8×8 determinant separated into two 4×4 blocks along the main diagonal.

H_{ij} formed from symmetric wave functions	0
0	H_{ij} formed from antisymmetric wave functions

$$= \begin{vmatrix} \text{Symmetric} \\ \text{block} \end{vmatrix} \times \begin{vmatrix} \text{Antisymmetric} \\ \text{block} \end{vmatrix} = 0$$

This factorization of the secular equation evidently permits us to consider the eigenvalues of the symmetric and antisymmetric wave functions separately. Take first the antisymmetric block.

Matrix elements here will be constructed from ψ_1^0, ψ_3^0, ψ_5^0, ψ_7^0. Some typical examples are

$$H_{11} = \tfrac{1}{2}\int [\psi_{1s}(1)\psi_{2s}(2) - \psi_{1s}(2)\psi_{2s}(1)]^2 f(1,2)\, d\tau \int \alpha^*(1)\alpha^*(2)\alpha(1)\alpha(2)\, d\omega$$

$$= \tfrac{1}{2}J + \tfrac{1}{2}J - K = J - K$$

$$H_{35} = \tfrac{1}{2}\iint \{[\psi_{1s}(1)\psi_{2s}(2)\alpha^*(1)\beta^*(2) - \psi_{1s}(2)\psi_{2s}(1)\alpha^*(2)\beta^*(1)]f(1,2)$$

$$\times [\psi_{1s}(1)\psi_{2s}(2)\beta(1)\alpha(2) - \psi_{1s}(2)\psi_{2s}(1)\beta(2)\alpha(1)]\}\, d\tau\, d\omega$$

$$= \tfrac{1}{2}\int \psi_{1s}(1)\psi_{2s}(2)f(1,2)\psi_{1s}(1)\psi_{2s}(2)\, d\tau \int \alpha^*(1)\beta^*(2)\beta(1)\alpha(2)\, d\omega$$

$$- \tfrac{1}{2}\int \psi_{1s}(1)\psi_{2s}(2)f(1,2)\psi_{1s}(2)\psi_{2s}(1)\, d\tau \int \alpha^*(1)\beta^*(2)\alpha(1)\beta(2)\, d\omega$$

$$- \tfrac{1}{2}\int \psi_{1s}(2)\psi_{2s}(1)f(1,2)\psi_{1s}(1)\psi_{2s}(2)\, d\tau \int \alpha^*(2)\beta^*(1)\beta(1)\alpha(2)\, d\omega$$

$$+ \tfrac{1}{2}\int \psi_{1s}(2)\psi_{2s}(1)f(1,2)\psi_{1s}(2)\psi_{2s}(1)\, d\tau \int \alpha^*(2)\beta^*(1)\beta(2)\alpha(1)\, d\omega$$

$$= 0 - \tfrac{1}{2}K - \tfrac{1}{2}K + 0 = -K$$

The first and last integrals in H_{35} vanish upon integration over the spin coordinates and there remains $H_{35} = -K$.

In a similar way the student should be able to confirm for himself the integrals $H_{11} = H_{77} = J - K$; $H_{33} = H_{55} = J$; $H_{13} = H_{15} = H_{17} = H_{37} = H_{57} = 0$. The secular equation for the antisymmetric block is thus

$$\begin{vmatrix} J-K-E & 0 & 0 & 0 \\ 0 & J-E & -K & 0 \\ 0 & -K & J-E & 0 \\ 0 & 0 & 0 & J-K-E \end{vmatrix} = 0$$

and we need its eigenvalues to find our perturbed energy levels. The roots of the secular equation are $J - K$, $J - K$, $J - K$, $J + K$, meaning that the degeneracy of the original antisymmetric levels is in part removed by the perturbing influence of Coulomb repulsion between the electrons. Upon deeper investigation, the integrals J and K turn out to be positive, so that the lower level is the triply degenerate root $J - K$ (called a triplet state) and the upper level is the nondegenerate root $J + K$ (called a singlet state). Upon investigation of the zero order wave functions constructed from the eigenvectors resulting from this secular equation, it turns out that the lower, triplet state corresponds to the configuration $(1s)(2s)$ with the electron spins "unpaired," meaning that the spins are parallel, either both up or both down. This is an illustration of a generalization which we shall later come to know as Hund's rule. Equivalently the singlet state is $(1s)(2s)$ with spins "paired"—one "up" the other "down."

We now turn our attention to the symmetric block which is a factor of the complete secular equation. The methods described above applied to this set of zero order wave functions lead to a secular equation,

$$\begin{vmatrix} J + K - E & 0 & 0 & 0 \\ 0 & J - E & K & 0 \\ 0 & K & J - E & 0 \\ 0 & 0 & 0 & J + K - E \end{vmatrix} = 0$$

with eigenvalues $J - K, J + K, J + K, J + K$, so that in this case the triplet level lies above the singlet. These results are summarized in the diagram of Figure 8.2. We see that the overall result of our perturbation calculation using all eight zero order wave functions for the configuration $(1s)(2s)$ is a splitting of the original octet level into two quartet levels. *This theoretical prediction fails miserably to explain the experimental results.* In fact the evidence from atomic spectroscopy is clear: the first excited state of the helium atom shows a splitting into a lower triplet and a somewhat higher singlet state with the other four states nowhere to be seen.

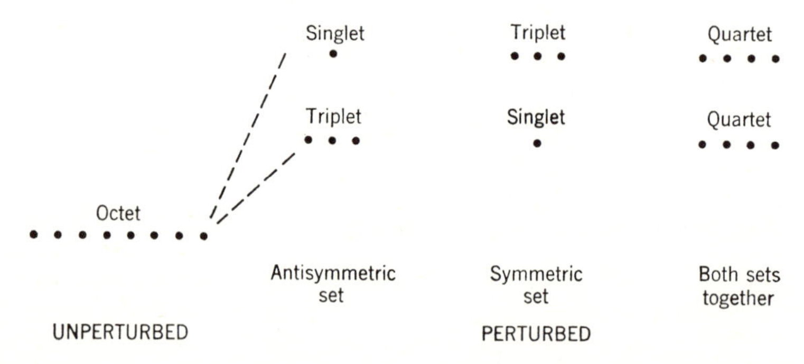

Figure 8.2 Splitting of the degenerate first excited state of the helium atom due to the perturbing influence of interelectronic repulsion.

PROBLEM

8.1 Zero order, orthonormal, antisymmetric wave functions for the $(1s)(2s)$ state of helium are

$$\psi_1 = \frac{1}{\sqrt{2}} \begin{vmatrix} \psi_{1s}(1)\alpha(1) & \psi_{1s}(2)\alpha(2) \\ \psi_{2s}(1)\alpha(1) & \psi_{2s}(2)\alpha(2) \end{vmatrix}$$

$$\psi_3 = \frac{1}{\sqrt{2}} \begin{vmatrix} \psi_{1s}(1)\alpha(1) & \psi_{1s}(2)\alpha(2) \\ \psi_{2s}(1)\beta(1) & \psi_{2s}(2)\beta(2) \end{vmatrix}$$

$$\psi_5 = \frac{1}{\sqrt{2}} \begin{vmatrix} \psi_{1s}(1)\beta(1) & \psi_{1s}(2)\beta(2) \\ \psi_{2s}(1)\alpha(1) & \psi_{2s}(2)\alpha(2) \end{vmatrix}$$

$$\psi_7 = \frac{1}{\sqrt{2}} \begin{vmatrix} \psi_{1s}(1)\beta(1) & \psi_{1s}(2)\beta(2) \\ \psi_{2s}(1)\beta(1) & \psi_{2s}(2)\beta(2) \end{vmatrix}$$

Define the Coulomb integral

$$J = \int [\psi_{1s}(1)\psi_{2s}(2)]^2 f(1, 2)\, d\tau$$

and the exchange integral

$$K = \int \psi_{1s}(1)\psi_{2s}(2) f(1, 2) \psi_{1s}(2)\psi_{2s}(1)\, d\tau$$

Express in terms of K and J the matrix elements H_{33}, H_{77}, H_{13} of the perturbation matrix

$$H_{ij} = \iint \psi_i^* f(1, 2) \psi_j \, d\tau \, d\omega$$

8.7 THE PAULI EXCLUSION PRINCIPLE

This evident contradiction between theory and experiment led Pauli to revise the foundations of quantum mechanics by the enunciation of his famous exclusion principle: the wave functions for a quantum mechanical system containing sets of identical particles must be either totally symmetric or totally antisymmetric with respect to every interchange of identical particles. Which set is to be chosen must be decided by an appeal to experiment, and for electrons the evidence is that the antisymmetric set is preferred. The student will have gathered by now that the methods of quantum mechanics involve the solution of a partial differential equation and the subsequent rejection of all solutions which do not satisfy certain auxiliary conditions. To date we have taken the auxiliary conditions to be that we require only those solutions which are finite, single valued, continuous, and normalized throughout the entirety of configuration space. According to the Pauli principle we must further reduce the set of acceptable wave functions by rejecting all those which do not have the right symmetry properties with respect to interchanges of identical particles. In other words: *exchange degeneracy does not exist.*

8.8 THE GROUND STATE OF LITHIUM

Armed with the Pauli principle, let us try to set up antisymmetric, independent electron wave functions for the ground state of lithium. By neglecting all interelectron repulsion terms $1/r_{ij}$ in the Hamiltonian, the Schrödinger equation for this three electron problem separates into three one electron, central field problems:

$$\begin{aligned} H^0(1, 2, 3) &= H(1) + H(2) + H(3) \\ \psi^0(1, 2, 3) &= \psi'(1)\psi''(2)\psi'''(3) \\ E^0 &= E' + E'' + E''' \end{aligned}$$

in which ψ', ψ'', ψ''' are any three solutions of the central field problem, $H(1)\psi'(1) = E'\psi'(1)$, and so on. We should most naturally expect that the independent electron configuration for the ground state is $(1s)^3$, corresponding to three electrons in the $1s$ level. Spin dependent wave functions are thus products of any three factors drawn from the rows

$$\begin{array}{ll} \psi_{1s}(1)\alpha(1); & \psi_{1s}(1)\beta(1) \\ \psi_{1s}(2)\alpha(2); & \psi_{1s}(2)\beta(2) \\ \psi_{1s}(3)\alpha(3); & \psi_{1s}(3)\beta(3) \end{array}$$

which is to say that for $\psi^0(1, 2, 3)$ we have the eight degenerate possibilities:

$$\psi_{1s}(1)\psi_{1s}(2)\psi_{1s}(3) \times \begin{cases} \alpha(1)\alpha(2)\alpha(3) & \uparrow\uparrow\uparrow \\ \alpha(1)\alpha(2)\beta(3) & \uparrow\uparrow\downarrow \\ \alpha(1)\beta(2)\alpha(3) & \uparrow\downarrow\uparrow \\ \alpha(1)\beta(2)\beta(3) & \uparrow\downarrow\downarrow \\ \beta(1)\alpha(2)\alpha(3) & \downarrow\uparrow\uparrow \\ \beta(1)\alpha(2)\beta(3) & \downarrow\uparrow\downarrow \\ \beta(1)\beta(2)\alpha(3) & \downarrow\downarrow\uparrow \\ \beta(1)\beta(2)\beta(3) & \downarrow\downarrow\downarrow \end{cases} \tag{8.7}$$

According to the Pauli principle, our next step is to organize these eight functions into eight orthonormal linear combinations which are either invariant or change only in sign when the coördinates of any two electrons are permuted. Of these eight linear combinations we should retain only the members of the antisymmetric set in our perturbation calculation. This program reaches a dead end when it turns out that *it is impossible to construct any antisymmetric linear combinations at all* from the set (8.7).

We shall demonstrate this fact by first constructing an antisymmetric, independent electron wave function for the known $(1s)^2(2s)$ ground state of lithium and then showing that the modifications in it required to construct the hypothetical $(1s)^3$ state force the atomic wave function to vanish. Write

$$\psi_1(1, 2, 3) = \frac{1}{\sqrt{3!}} \begin{vmatrix} \psi_{1s}(1)\alpha(1) & \psi_{1s}(2)\alpha(2) & \psi_{1s}(3)\alpha(3) \\ \psi_{1s}(1)\beta(1) & \psi_{1s}(2)\beta(2) & \psi_{1s}(3)\beta(3) \\ \psi_{2s}(1)\alpha(1) & \psi_{2s}(2)\alpha(2) & \psi_{2s}(3)\alpha(3) \end{vmatrix}$$

When expanded, this determinant proves to be a linear combination of products of one electron functions. It is antisymmetric upon interchange of any two electron coordinates, for if we permute coordinates 1 and 2, then the new determinant is derived from the old by interchanging the first two columns, and a known property of determinants is that the interchange of any two columns or any two rows simply changes the algebraic sign of the determinant. The quantity $\sqrt{3!}$ appears as a normalizing factor for the $3! = 6$ terms in the linear combination.

Were we to use this recipe to construct the $(1s)^3$ state, then the ψ_{2s} functions in the last row would be replaced by ψ_{1s} functions, and the third and first rows would be identical. A determinant with two identical rows or columns is always zero: hence the $(1s)^3$ state does not exist.

A second determinant $\psi_2(1, 2, 3)$ for the lithium $(1s)^2(2s)$ state can be constructed by writing β spin functions along the last row instead of α spin functions. It too will be antisymmetric upon exchange of any pair of electron coordinates, and like $\psi_1(1, 2, 3)$, it will vanish if we attempt to write a $(1s)^3$ state by replacing ψ_{2s} functions with ψ_{1s}. It follows that, like hydrogen, the ground state wave function of the lithium atom is doubly degenerate due to the two possible orientations of the spin of the valence electron.

With the example of lithium in mind, we should also be prepared to accept the alternative and more familiar statement of the Pauli principle: no pair of electrons in an atom or molecule can occupy states having identical quantum numbers including spin. The $(1s)^3$ configuration in lithium is thus prohibited, for two electrons in the $1s$ level must have spins opposed $\uparrow\downarrow$ after the manner of the ground state wave function (8.5) for helium, and a third electron in the $1s$ level would have to have a spin matching one of those already present.

8.9 THE AUFBAU PRINZIP AND THE PERIODIC TABLE

A knowledge of the central field problem together with the hypothesis of electron spin and the Pauli principle permits a construction of the periodic table. A solution ignoring spin of the one electron central field problem using a radial potential which approximates to the attraction of the nucleus plus the averaged effect of all the other electrons provides us with a set of one electron wave functions or *atomic orbitals*. These orbitals will have angular factors and angular quantum numbers l and m the same as the atomic orbitals of hydrogen, but the radial factors will be different from those of hydrogen; and this latter consideration also leads to a dependence of the energy levels on two quantum numbers l and n rather than on n alone as for hydrogen. The angular quantum numbers l and m have the same restrictions as for hydrogen: $l = 0, 1, 2, \ldots$; $m = -l, -l + 1, \ldots, 0, \ldots, l - 1, l$ but the quantum numbers n are no longer necessarily integers. It is nevertheless conventional to refer an atomic orbital to the corresponding quantum number for hydrogen, so that atomic orbitals are labeled $1s$, $2s$, $2p, \ldots$, and so on, despite the fact that in a non-Coulomb central field the principal quantum number n no longer takes on values $1, 2, \ldots$. The restriction $l = n - 1$ where n is the integral principal quantum number of hydrogen is, however, still valid. The effect of using a non-Coulomb central field for the purpose of constructing one electron atomic orbitals has the effect of removing some of the degeneracy involved in the hydrogen-like orbitals, so that s, p, d, etc., orbitals having the same principal quantum number n no longer have the same energy. This situation is summarized in Figure 8.3. The reader should note for the lower levels at least, that the energy is still more sensitive to n than to l. For the higher levels, this ceases to be the case, and in some atoms $4s$ is actually lower in energy than $3d$.

Once the energy levels and their associated one electron wave functions ψ_{1s}, ψ_{2s}, and so on, have been determined, we use these atomic orbitals multiplied by spin functions α and β to construct zero order, independent electron wave functions for a many electron atom. All terms in the many electron Hamiltonian which involve interaction of the electrons are neglected in the zeroth order, and the approximate energy levels of the real atom are calculated by bringing in the interaction terms as a perturbation.

A convenient recipe for cooking the zero order wave function for the ground state of a many electron atom is the Aufbau Prinzip of Bohr. Starting with a basket of protons and neutrons and another basket of electrons, we

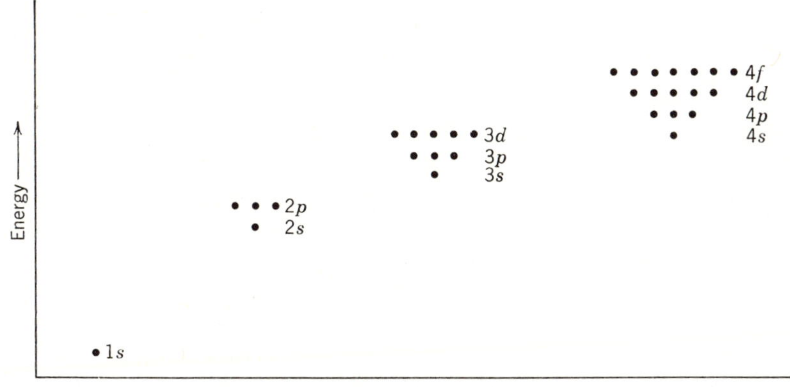

Figure 8.3 Single electron energy levels in a non-Coulomb central field (schematic).

imagine the gradual construction of heavier and heavier atoms by mixing appropriate quantities of each particle in a casserole. To the zeroth order, the electrons do not interact with each other, so that the Pauli principle is satisfied simply by assigning them to successively higher energy states taking care that spin quantum numbers are also assigned such that no two electrons have all four quantum numbers the same.

For hydrogen this leads to two possible ground state wave functions $(1s)\alpha$ and $(1s)\beta$. For helium we have demonstrated the existence of a single ground state wave function $(1s)^2\alpha\beta$, and this completes the $1s$ shell; for no further electrons can be accommodated in $1s$ without violating the Pauli principle. Lithium marks the first appearance of a $2s$ electron superimposed upon the helium-like closed shell of two $1s$ electrons, and for lithium, as discussed in Section 8.8, there are two ground state wave functions $(\mathrm{He})(2s)\alpha$ and $(\mathrm{He})(2s)\beta$. With beryllium the $2s$ shell is also closed, with boron the first $2p$ electron appears. At this point the number of possibilities for a ground state wave function becomes greater, for the $2p$ electron can exist in six distinct states of equivalent zero order energy: $(2p_{-1})\alpha$, $(2p_{-1})\beta$, $(2p_0)\alpha$, and so on. For carbon with two $2p$ electrons there turn out to be 15 ways in which different orbital and spin assignments can be made.

We suppose the *aufbau* to continue in this way so that quantum numbers are assigned to electrons in such a way that no two electrons have all four quantum numbers identical and that all the lower lying orbitals are filled. As we have seen, this is often likely to lead to a number of equivalent possibilities all of which have the same zero order ground state energy. To obtain the energy of the atom to first order, all ground state wave functions of equal energy must be used in a perturbation calculation; and generally such a calculation shows a splitting of the zero order ground state energy into a number of terms. We have already experienced an example of this splitting in our calculation of the first excited state of helium where we showed that the four, degenerate, antisymmetric, zero order wave functions for the configuration $(1s)(2s)$ split upon perturbation with the interelectronic interaction potential $f(1, 2)$ into two levels, a lower triplet and an upper singlet. Were we to do the same for the configuration $(\text{Be})(2p)^2$ (beryllium core plus two $2p$ electrons = carbon), all fifteen ground state wave functions would have to appear in our perturbation calculation, leading to a secular equation of order 15×15. This formidable proposition can fortunately be simplified by an appeal to the angular momentum and spin operators, but I avoid the complicated details.[5–7] The final results are that the fifteen fold degenerate zero order ground state of carbon splits upon perturbation with interelectronic repulsion operators into three terms, the lowest of which is ninefold degenerate, the next highest fivefold degenerate, and the highest nondegenerate.

8.10 HUND'S RULE

The results of calculations of this sort as well as experimental evidence from atomic spectroscopy have led to a generalization known as Hund's rule. From the point of view of orbital and spin assignments to electrons in a zero order atomic wave function, Hund's rule states that due to the splitting of the energy levels because of interelectronic interaction, those zero order wave functions are favored which possess the maximum number of unpaired electrons. Thus the two $2p$ electrons of carbon will have their spins parallel, and by the Pauli principle they must therefore reside in different $2p$ orbital functions; that is, they must have different assignments of the quantum number m. Similarly the nitrogen atom with its three $2p$ electrons will have each of them assigned to a different $2p$ orbital, so that the three $2p$ orbitals of nitrogen contain 1 electron each. Passing on to oxygen, the fourth $2p$ electron must pair its spin with one of the others, leaving the remaining two electrons in different $2p$ orbitals with spins unpaired. The addition of two more $2p$ electrons brings us to neon, and the $2p$ shell is complete.

The aufbau continues in a regular way until we get to potassium. Building upon the argon core $(1s)^2(2s)^2(2p)^6(3s)^2(3p)^6$, in the normal scheme of things we might expect the next electron to go into the $3d$ level. The influence of the non-Coulomb part of our central field is by now so large, however, that the $4s$ level lies slightly lower than the $3d$, so that potassium is actually $(\text{Ar})(4s)$. Of course, the expression "non-Coulomb central field" is a fictitious artifice on our part, designed to absorb as much as possible of the interelectronic interaction in an atom without destroying the computational convenience of single electron, central field wave functions. A more physically realistic statement would be to say that the $4s$ level lies lower than the $3d$ level in potassium because of interelectronic repulsion.

The results of *aufbau* of this sort have led to lists of ground state electronic assignments to all the atoms in the periodic table, at least up to element 100, fermium.[8] Such tables, and the inversion of the order of hydrogen-like energy levels such as $4s$ and $3d$ due to interelectronic interaction, have led to satisfactory explanations of the existence of the transition metals and the rare earths.

8.11 SPIN-ORBIT INTERACTION

The perturbation of the ground state zero order energies of atoms due to Coulomb repulsion between the electrons has led, as we have seen, to a splitting of the originally degenerate levels into different terms. Often these terms are themselves degenerate, but the evidence from atomic spectroscopy is such as to show a further fine splitting into closely spaced levels. This phenomenon is explained by bringing a second perturbing term into the Hamiltonian, that of the interaction of the magnetic field due to the spin of the electrons and the magnetic field due to their orbital rotation. Other than to recognize its existence, this topic lies hopelessly outside the scope of this book.

8.12 LIMITATIONS OF THE ORBITAL PICTURE

All of our perturbation calculations in this chapter and indeed much of the language of atomic physics and chemistry has been based upon the assumption that the independent electron model is a reasonably valid starting point for the quantum mechanical description of the electronic wave function of an atom. An enormous body of experimental evidence supports the usefulness of this approach, and shop talk between chemists of widely different interests and training is likely to be sprinkled with references to "*d* electrons", "unpaired spins," and so on. However picturesque, such language carries with it the danger that the user will perhaps accept the independent electron model too uncritically; for careful and exhaustive calculations have shown that there are inherent limitations in the orbital picture.

The independent electron model inevitably leads to an approximate electronic wave function written as a linear combination of products of one electron "orbital" functions, usually having a hydrogen-like or central field character. Such orbital functions suffer from the defect that they deemphasize the instantaneous correlations between the motions of the electrons. This defect is minimized in the Hartree-Fock method where the nature of the central field is adjusted to minimize the error, but the error is never completely eliminated.

In order to obtain accurate theoretical predictions of a quality comparable to the accuracy of the available experimental results, it has proved necessary to abandon the use of orbital wave functions and to take explicit account of the instantaneous correlation in the electronic motion due to the presence in the Hamiltonian of the electronic repulsion terms. Such calculations are invariably based upon the variation method to be introduced in the next chapter and will thus not be discussed here. It is sufficient to say that the most accurate calculations on many electron atoms and molecules have been obtained by abandoning from the outset any suggestion that the electrons move independently. In such calculations there is little justification for labeling electrons as $1s$ or $3p$, and so on, and in the computations there arise no such quantities as Coulomb or exchange integrals. The persistence of this language in our vocabulary should thus be taken with some degree of sophisticated tolerance for the limitations involved.

PROBLEM

8.2 Referring to Problem 4.7, assume that two electrons in a closed ring move independently of each other without mutual electrostatic repulsion. Write down properly antisymmetrized wave functions for this two electron problem.

(1) For the ground state.

(2) For the first excited state.

REFERENCES

1. H. Eyring, J. Walter, and G. E. Kimball, *Quantum Chemistry*, John Wiley and Sons Inc., New York, 1944, pp. 369–371.
2. H. Eyring, J. Walter, and G. E. Kimball, *ibid.*, pp. 163–167.
3. W. Kauzmann, *Quantum Chemistry*, Academic Press, New York, 1957, pp. 334–336.
4. F. L. Pilar, *Elementary Quantum Chemistry*, McGraw-Hill, Inc., New York, 1968. Chapter 13, pp. 186–193.
5. H. Eyring, J. Walter, and G. E. Kimball, *loc. cit.*, pp. 140–143.
6. W. Kauzmann, *loc. cit.*, pp. 337–344.
7. F. L. Pilar, *loc. cit.*, Chapter 12.
8. L. Pauling, *The Nature of the Chemical Bond*, Cornell University Press, 1960.

Chapter 9

THE VARIATION METHOD

9.1 INTRODUCTION

Perturbation methods have enabled us to build up a semiquantitative picture of atomic structure. Similar in its computational details but different in its logical approach to problems in quantum mechanics is the variation method—another source of approximate results based on the Schrödinger equation and the only one that has led to a useful picture of the chemical bond. We start our study of it with an intuitive examination of a variation problem in the vector space of two dimensions.

9.2 AN ILLUSTRATIVE EXAMPLE

Consider the function

$$Q(x_1, x_2) = \lambda_1 x_1^2 + \lambda_2 x_2^2$$

and plot in the x_1x_2 plane the contours $Q =$ constant (Figure 9.1). For λ_1 and λ_2 both positive, these contour lines have the form of ellipses centered on the origin, and the contour plot may be interpreted as a valley in that the altitude Q increases as one moves away from the origin in any direction. Now superimpose on the Q contours the unit circle

$$x_1^2 + x_2^2 = 1$$

(dotted line) and consider the changes in Q experienced by a pedestrian as he traverses the circle. Starting on the x_1 axis and moving counterclockwise, his altitude evidently increases from contour 2 to contour 3 and then decreases again. In fact, these contours define the minimum and the maximum altitudes attained so long as the pedestrian does not leave the unit circle. In mathematical language, the altitude Q is said to be stationary with respect to the

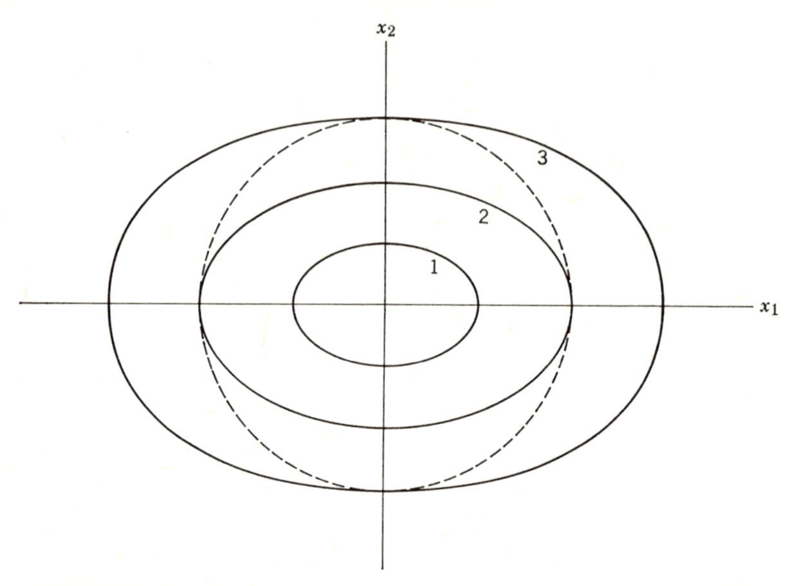

Figure 9.1 Geometric interpretation of a variation problem.

restriction $x_1^2 + x_2^2 = 1$ whenever the pedestrian crosses the x_1 or x_2 axes; for an infinitesimal displacement along the unit circle at these points neither increases nor decreases Q.

In vector language the problem may be stated as follows. Given a matrix

$$H = \begin{pmatrix} \lambda_1 & 0 \\ 0 & \lambda_2 \end{pmatrix}$$

find all vectors $\mathbf{x} = (x_1, x_2)$ such that $Q = \mathbf{x}^T H\mathbf{x}$ is stationary under the restriction $\mathbf{x}^T\mathbf{x} = 1$. Our knowledge of vector algebra is by now sufficiently refined that we may observe that the required vectors are two in number and that they are the normalized eigenvectors of H: $\mathbf{x} = \mathbf{e}_1 = (1, 0)$ and $\mathbf{x} = \mathbf{e}_2 = (0, 1)$. Furthermore, when $\mathbf{x} = \mathbf{e}_1$, the value of Q at this stationary point is $Q(1, 0) = \lambda_1$; and when $\mathbf{x} = \mathbf{e}_2$, the value of Q at this second stationary point is $Q(0, 1) = \lambda_2$. There is evidently a close relationship between the stationary points of what is known as a quadratic form

$$\begin{aligned} Q(x_1, x_2, \ldots, x_n) &= \sum \sum h_{ij} x_i x_j \\ &= \mathbf{x}^T H \mathbf{x} \end{aligned}$$

and the eigenvectors of the symmetric matrix H when admissible stationary points are restricted to be the termini of unit vectors $\mathbf{x}^T\mathbf{x} = \sum \mathbf{x}_i^2 = 1$. A general theorem (which will not be proved here) states that the two problems are, in fact, identical.[1]

9.3 ANOTHER ILLUSTRATIVE EXAMPLE

Let us push our geometric intuitions to a problem in the vector space of three dimensions. Suppose that we have a symmetric matrix H each of whose eigenvalues is positive and distinct (nondegenerate). From it we construct a quadratic form

$$Q(x_1, x_2, x_3) = \mathbf{x}^T H \mathbf{x}$$

where $\mathbf{x} = (x_1, x_2, x_3)$ is any vector in the space. Now consider the succession of concentric ellipsoidal surfaces over each of which Q assumes some constant value. It may be proved that the surfaces Q = constant are indeed ellipsoids whenever the eigenvalues of H are positive, and that in addition the eigenvectors of H lie along the principal axes of these ellipsoids. Superimpose upon these ellipsoids the unit sphere $\mathbf{x}^T\mathbf{x} = 1$ and consider the variations of Q as an astronaut wanders around the surface of the sphere. It does not require much imagination to generalize the situation from that of Section 9.2 and to expect what is actually the case—that the stationary points are the termini of the normalized eigenvectors of H, and that Q at these stationary points takes on the successive values λ_1, λ_2, λ_3, a maximum, a minimum, and an intermediate inflection value.

An exact calculation of these stationary points would require the solution of a secular equation of order 3, but if we are interested primarily in the minimum eigenvalue and are willing to accept an approximate result, we can

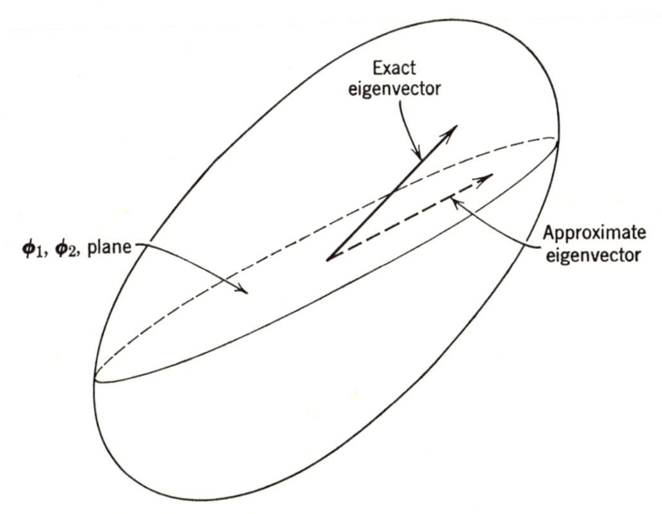

Figure 9.2 Relation between an approximate and an exact eigenvector in the variation method.

reduce the order of the secular equation by the following device. We *assume* that the eigenvector corresponding to the minimum eigenvalue can be expressed as a linear combination of *two* vectors,

$$\mathbf{x} = c_1\boldsymbol{\phi}_1 + c_2\boldsymbol{\phi}_2$$

where the vectors $\boldsymbol{\phi}_1$ and $\boldsymbol{\phi}_2$ are selected on intuitive or other grounds by the investigator. Geometrically this amounts to imposing upon the problem an additional restriction: the desired eigenvector is presumed to lie within the plane spanned by $\boldsymbol{\phi}_1$ and $\boldsymbol{\phi}_2$, and in Figure 9.2 I have drawn such an intersecting plane. If the bold face line represents a true eigenvector, then a calculation based upon the assumption stated above will yield a solution represented by the dotted vector—a projection of the true eigenvector into the plane. Let us see how this might come about by a sample calculation.

9.4 ARITHMETIC DETAILS

Given a matrix

$$H = \begin{pmatrix} 2 & 1 & 0 \\ 1 & 2 & 1 \\ 0 & 1 & 2 \end{pmatrix}$$

find approximately its minimum eigenvalue and the corresponding eigenvector.

By standard methods the student may determine the exact eigenvalues and eigenvectors of H to be

$$\begin{aligned} \lambda_1 &= 2 - \sqrt{2} = 0.586; & \mathbf{E}_1 &= (\tfrac{1}{2}, -1/\sqrt{2}, \tfrac{1}{2}) \\ \lambda_2 &= 2; & \mathbf{E}_2 &= (1/\sqrt{2}, 0, -1/\sqrt{2}) \\ \lambda_3 &= 2 + \sqrt{2} = 3.414; & \mathbf{E}_3 &= (\tfrac{1}{2}, 1/\sqrt{2}, \tfrac{1}{2}) \end{aligned}$$

Now let us obtain an approximation to the lowest eigenvalue and its eigenvector by assuming that it can be expressed as a linear combination of two orthonormal basis vectors

$$\mathbf{E}_1 \sim \mathbf{x} = c_1\boldsymbol{\phi}_1 + c_2\boldsymbol{\phi}_2 \tag{9.1}$$

where

$$\boldsymbol{\phi}_1 = (1, 0, 0)$$
$$\boldsymbol{\phi}_2 = (0, 1, 0)$$

For $\mathbf{x}$ to be normalized we must have $c_1^2 + c_2^2 = 1$, but the c's are otherwise unspecified, and indeed it is our task to vary them so as to make $\mathbf{x}$ the best approximation to $\mathbf{E}_1$ which is possible under the circumstances. From our general theorem, we know that the eigenvectors of H have the property that they make the quadratic form

$$Q = \mathbf{x}^T H \mathbf{x} \tag{9.2}$$

stationary whenever $\mathbf{x}$ is one of the eigenvectors of H. Substituting from (9.1) into (9.2),

$$\begin{aligned} Q = \mathbf{x}^T H \mathbf{x} &= (c_1 \boldsymbol{\phi}_1 + c_2 \boldsymbol{\phi}_2)^T H (c_1 \boldsymbol{\phi}_1 + c_2 \boldsymbol{\phi}_2) \\ &= c_1^2 H_{11} + c_1 c_2 H_{12} + c_2 c_1 H_{21} + c_2^2 H_{22} \end{aligned} \tag{9.3}$$

in which $H_{ij} = \boldsymbol{\phi}_i^T H \boldsymbol{\phi}_j$,

$$H_{11} = (1, 0, 0) \begin{pmatrix} 2 & 1 & 0 \\ 1 & 2 & 1 \\ 0 & 1 & 2 \end{pmatrix} \begin{pmatrix} 1 \\ 0 \\ 0 \end{pmatrix} = 2$$

$$H_{22} = (0, 1, 0) \begin{pmatrix} 2 & 1 & 0 \\ 1 & 2 & 1 \\ 0 & 1 & 2 \end{pmatrix} \begin{pmatrix} 0 \\ 1 \\ 0 \end{pmatrix} = 2$$

$$H_{12} = (1, 0, 0) \begin{pmatrix} 2 & 1 & 0 \\ 1 & 2 & 1 \\ 0 & 1 & 2 \end{pmatrix} \begin{pmatrix} 0 \\ 1 \\ 0 \end{pmatrix} = 1 = H_{21}$$

Expressed in terms of the c's, the quadratic form (9.3) is thus

$$Q = 2c_1^2 + c_1 c_2 + c_2 c_1 + 2c_2^2 = 2c_1^2 + 2c_1 c_2 + 2c_2^2$$

But to this approximation Q is now a quadratic form in the c's with a two dimensional matrix

$$K = \begin{pmatrix} 2 & 1 \\ 1 & 2 \end{pmatrix}$$

and our problem is still to make Q stationary by varying the c's under the restriction $c_1^2 + c_2^2 = 1$. We solve this problem by reverting to eigenvalue-eigenvector formalism based upon the matrix K.

The student will perceive that the approximation of taking the three dimensional eigenvectors of H to be linear combinations of two arbitrarily chosen orthonormal vectors $\boldsymbol{\phi}_1$ and $\boldsymbol{\phi}_2$ has resulted in a reduction in the size of the problem from a secular equation of order 3 to one of order 2:

$$|K - \lambda I| = \begin{vmatrix} 2 - \lambda & 1 \\ 1 & 2 - \lambda \end{vmatrix} = 0$$

The price we pay for this reduction in size is a loss in accuracy of the computed eigenvalue and eigenvector.

For the new matrix K we find easily

$$\lambda_1 = 1; \qquad \boldsymbol{\varepsilon}_1 = \left(\frac{1}{\sqrt{2}}, -\frac{1}{\sqrt{2}}\right)$$

$$\lambda_2 = 3; \qquad \boldsymbol{\varepsilon}_2 = \left(\frac{1}{\sqrt{2}}, \frac{1}{\sqrt{2}}\right)$$

and the components of these two dimensional eigenvectors are the coefficients c_1, c_2 which must be used together with the basis vectors $\boldsymbol{\phi}_1$, $\boldsymbol{\phi}_2$ to construct approximate eigenvectors for H. As an example, for the lowest eigenvalue we have $\lambda_1 \sim 1$ and $\mathbf{E}_1 \sim (1/\sqrt{2})\boldsymbol{\phi}_1 - (1/\sqrt{2})\boldsymbol{\phi}_2 = (1/\sqrt{2}, -1/\sqrt{2}, 0)$. These should be compared with the exact values $\lambda_1 = 0.586$ and $\mathbf{E}_1 = (\frac{1}{2}, -1/\sqrt{2}, \frac{1}{2})$. It is evident that our choice of the basis $\boldsymbol{\phi}_1$ and $\boldsymbol{\phi}_2$ has led to a considerable error.

PROBLEM

9.1 Repeat the calculation of Section 9.4 using as a basis

$$\boldsymbol{\phi}_1 = (1/\sqrt{2}, 0, 1/\sqrt{2})$$
$$\boldsymbol{\phi}_2 = (0, 1, 0)$$

9.5 THE VARIATION THEOREM

Probably you skipped Problem 9.1. If not, you found that the basis chosen there was superior to that of Section 9.4 in that it yielded an exact value to the minimum eigenvalue. Geometrically this means that we were so lucky in Problem 9.1 as to choose a basis in whose plane happened to lie the minimal eigenvector $\mathbf{E}_1$ of H. One is rarely so lucky in quantum mechanical investigations, but the comparative success of variation calculations may be assessed by an important theorem. The true least eigenvalue of a symmetric matrix is always less than or equal to any approximate eigenvalue calculated from a variation approximation. If, therefore, in a series of variation calculations we obtain a succession of approximate least eigenvalues, that approximation is best which is lowest, and its corresponding eigenvector is the most accurate approximation to the true minimal eigenvector.

Parenthetically, it is not inappropriate to remark that our preoccupation with the minimum eigenvalue is due to our primary interest in applications to the ground states of atomic and molecular systems. The higher eigenvalues and eigenvectors obtained by variation approximations constitute estimates of the higher energy levels and eigenstates of the system in hand, but the approximation generally becomes increasingly poor the higher the eigenvalue.

As examples of these remarks, the student should compare the exact and approximate results obtained in Section 9.4 for which we have approximately $\lambda_1 \sim 1 > 0.586$.

PROBLEM

9.2 (1) Determine the exact eigenvalues and eigenvectors of the matrix

$$H = \begin{pmatrix} 1 & 1 & 0 \\ 1 & 1 & 1 \\ 0 & 1 & 1 \end{pmatrix}$$

(2) Select any two orthonormal, three dimensional basis vectors of your choice, and obtain by the variational method an approximation to the lowest eigenvalue of H and its eigenvector.

9.6 A GENERALIZED VARIATION PROBLEM

Often in practice the variation problem presented in a quantum mechanical investigation takes a more general form: find the stationary points of a quadratic form

$$Q = \mathbf{x}^T H \mathbf{x}$$

subject to the restriction

$$F = \mathbf{x}^T S \mathbf{x} = 1$$

where F is some other quadratic form. If the symmetric matrix S is the unit matrix $S = I$, then we have the problem of Sections 9.2 to 9.4. If $S \neq I$, then the contour $F = 1$ in the special case of a two dimensional vector space is some other ellipse, and the track our pedestrian is asked to follow would have the general appearance of Figure 9.3.

In the diagram I have placed an asterisk near the stationary points where the contours $Q =$ constant and $F = 1$ are locally tangent and Q does not change for an infinitesimal displacement along the F contour. It seems intuitively acceptable that these stationary points lie along two vectors, but the vectors are neither the eigenvectors of H nor of S, nor do they constitute an orthonormal set. In an n dimensional vector space it can be shown that the

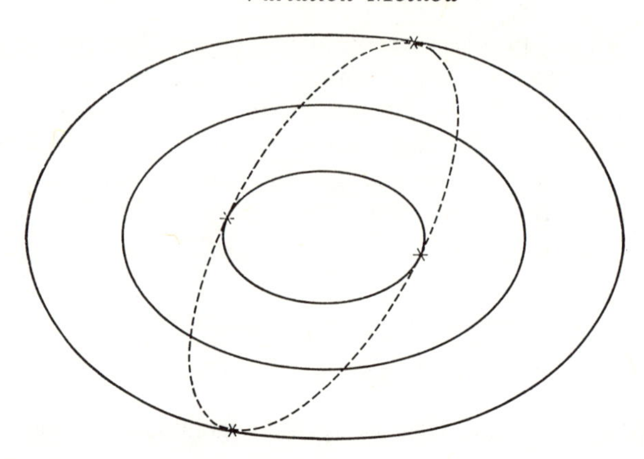

Figure 9.3 Geometric interpretation of a generalized variation problem.

stationary points do indeed lie at the termini of the n linearly independent vector solutions $\mathbf{x} = (x_1, x_2, \ldots, x_n)$ of the generalized eigenvalue problem $H\mathbf{x} = \lambda S\mathbf{x}$ whose secular equation is

$$|H - \lambda S| = 0$$

We shall see in practice that this type of secular equation arises whenever we use for a variation calculation a set of nonorthonormal basis vectors.

9.7 A SAMPLE CALCULATION

Given

$$H = \begin{pmatrix} 5 & 2 \\ 2 & 2 \end{pmatrix} \quad \text{and} \quad S = \begin{pmatrix} 1 & \frac{1}{2} \\ \frac{1}{2} & 1 \end{pmatrix}$$

find the stationary points of $Q = \mathbf{x}^T H\mathbf{x}$ subject to the restriction $F = \mathbf{x}^T S\mathbf{x} = 1$. We first set up the secular equation

$$\begin{vmatrix} 5 - \lambda & 2 - \frac{1}{2}\lambda \\ 2 - \frac{1}{2}\lambda & 2 - \lambda \end{vmatrix} = 0$$

The roots of the characteristic polynomial prove to be $\lambda_1 = 5.097$; $\lambda_2 = 1.570$. The components of the first eigenvector satisfy the simultaneous, homogeneous equations

$$(5 - 5.097)x_1 + (2 - (\tfrac{1}{2})(5.097))x_2 = 0$$
$$(2 - (\tfrac{1}{2})(5.097))x_1 + (2 - 5.097)x_2 = 0$$

whence we find an unnormalized first eigenvector (1, −0.177). To normalize it, recall that for this problem we do not require $\mathbf{x}^T\mathbf{x} = 1$ but rather $\mathbf{x}^T S\mathbf{x} = 1$. Compute

$$(1, -0.177)\begin{pmatrix} 1 & \frac{1}{2} \\ \frac{1}{2} & 1 \end{pmatrix}\begin{pmatrix} 1 \\ -0.177 \end{pmatrix} = 0.854$$

Hence if we divide through each component of the unnormalized eigenvector by $(0.854)^{1/2} = 0.924$, we have an eigenvector "normalized" according to the special requirements of the problem,

$$\mathbf{E}_1 = (1.082, -0.192)$$

In a similar way the second eigenvector correponding to $\lambda = 1.570$ and normalized as above turns out to be

$$\mathbf{E}_2 = (0.379, -1.069)$$

Note that these eigenvectors are not orthogonal, $\mathbf{E}_1{}^T\mathbf{E}_2 \neq 0$. The student will find, however, that they do satisfy the rule $\mathbf{E}_1{}^T S\mathbf{E}_2 = 0$, so that the eigenvectors obey the general law

$$\mathbf{E}_i{}^T S\mathbf{E}_j = \delta_{ij}$$

PROBLEM

9.3 Find the stationary points of $Q = \mathbf{x}^T H\mathbf{x}$ subject to $\mathbf{x}^T S\mathbf{x} = 1$ where

$$H = \begin{pmatrix} 3 & 1 \\ 1 & 2 \end{pmatrix} \quad \text{and} \quad S = \begin{pmatrix} 1 & 1 \\ 1 & 2 \end{pmatrix}$$

9.8 FIRST ORDER PERTURBATION THEORY AND THE VARIATION METHOD

First order perturbation theory is also founded upon writing approximate eigenvectors for an operator H by using as a basis a finite set of eigenvectors of some other, simpler operator H^0. The end result, as we have seen, is always a secular equation of finite size. The relation between this approach and the more general variation method is simply that when appropriate, we let our intuitive choice of a basis for a variation calculation be guided by the fact that the new eigenvalues and eigenvectors of the exact operator H must differ only slightly from those of the simplified operator H^0. Degenerate perturbation theory assumes as a basis all the zero order eigenvectors of an unperturbed, degenerate eigenvalue. According to the variation method, better approximate eigenvalues and eigenvectors can be obtained if the basis chosen includes not only the zero order eigenvectors of the degenerate level, but also the zero order eigenvectors of closely lying, neighboring eigenvalues. The resulting secular equation will, of course, be larger; but that is the price to be paid for greater accuracy. We shall see the results of this "mixing" of the zero order eigenvectors of closely spaced eigenvalues in the hybridization and resonance phenomena of chemical bond theory.

9.9 THE GENERALIZATION TO FUNCTION SPACES

The pictures and language which we have adopted so far are those of finite dimensional vector spaces. The procedure needed to carry out a variation calculation for an Hermitian operator is based upon the variation theorem of quantum mechanics: the Schrödinger equation is equivalent to finding all functions ψ which make the integral

$$E = \int \psi^* H \psi \, d\tau$$

stationary under the restriction

$$\int \psi^* \psi \, d\tau = 1$$

By expressing ψ as a linear combination of *any* infinite set of linearly independent functions ϕ defined in the configuration space of the system, this problem could in principle be solved exactly whether or not the ϕ's were eigenfunctions of H. In practice, however, an infinite set of basis functions would lead to a secular equation of infinite order, so that to make the problem tractable, we are forced to assume that the ground state of the system can be represented approximately by a finite linear combination of basis functions $\phi_1, \phi_2, \ldots, \phi_n$. That is,

$$\psi = c_1\phi_1 + c_2\phi_2 + \cdots + c_n\phi_n \tag{9.4}$$

where the ϕ's are selected according to the ingenuity of the investigator and the c's are to be determined. By substituting this finite linear combination into the variation integral, we find that the quadratic form

$$E \sim Q = \sum_1^n \sum_1^n H_{ij} c_i c_j$$

is to be made stationary, but that normalization of ψ requires the restriction

$$\sum \sum S_{ij} c_i c_j = 1$$

Here the H_{ij} and the S_{ij} are integrals

$$H_{ij} = \int \phi_i^* H \phi_j \, d\tau$$

$$S_{ij} = \int \phi_i^* \phi_j \, d\tau$$

By the device of adopting a finite set of basis functions, we have thus reduced the problem to manageable proportions, but the price we pay for this simplification is to limit the possible eigenfunctions to an n dimensional subspace of the infinite dimensional function space of the Hamiltonian operator. Our approximate eigenvectors are thus only projections of the exact eigenfunctions upon this subspace and are no better than our skill in guessing an accurate basis set of ϕ's in which to express the ground state of the system.

The secular equation whose lowest eigenvalue approximates but is never less than the ground state energy of the system is

$$|H_{ij} - ES_{ij}| = 0 \tag{9.5}$$

If the ϕ_i constitute an orthonormal set, then $S_{ij} = \delta_{ij}$, and the secular equation is of the type by now familiar to us. Otherwise the procedure of Sections 9.6 and 9.7 must be followed.

The higher eigenvalues of equation 9.5 are approximations to the higher energy levels of the system, but the approximation becomes increasingly poor the higher the eigenvalue. In any case, the eigenvectors $\mathbf{c} = (c_1, c_2, \ldots, c_n)$ of the secular equation 9.5 (which I shall for simplicity assume to be real) are the sets of coefficients appearing in the expansions

$$\psi = c_1\phi_1 + c_2\phi_2 + \cdots + c_n\phi_n$$

and these expansions are approximate wave functions for the system. The c's must be normalized according to the rule

$$\mathbf{c}^T S \mathbf{c} = \sum \sum S_{ij} c_i c_j = 1 \tag{9.6}$$

and if $S_{ij} = \delta_{ij}$, equations 9.5 and 9.6 reduce to the familiar

$$|H_{ij} - E\delta_{ij}| = 0$$
$$c_1^2 + c_2^2 + \cdots + c_n^2 = 1 \tag{9.7}$$

REFERENCES

1. R. Courant and D. Hilbert, *Methods of Mathematical Physics*, Vol. 1, Interscience, New York, 1953, pp. 23–26.
2. H. Eyring, J. Walter, and G. E. Kimball, *Quantum Chemistry*, John Wiley and Sons, Inc., New York, 1944, pp. 99–101.
3 L. Pauling and E. B. Wilson, Jr., *Introduction to Quantum Mechanics*, McGraw-Hill, Inc., New York, 1935, Chapter 7.
4. W. Kauzmann, *Quantum Chemistry*, Academic Press, New York, 1957, pp. 119–20.

Chapter 10

THE DELTA WORLD

10.1 INTRODUCTION

It is tantalizing for the chemist to have in his possession a completely reliable mathematical theory that presumably describes in detail the internal mechanics of atoms and molecules and to be unable to carry that theory to computational completion. It is a lamentable fact, however, that even the impressive advances in computing machinery witnessed by the past two decades have been insufficient to unlock the ultimate treasures that the Schrödinger theory holds for the chemist in the form of accurate predictions of the energy levels and geometries of such experimentally short lived species as the methyl radical or most of the transition state complexes. To the extent that quantum mechanics is to be at all useful in chemistry, we are therefore forced to rely on brutally crude approximation methods, and much of the rest of this book will be concerned with the intuitive construction of qualitative quantum mechanical procedures which will help interpret the grosser features of molecular behavior. A great deal of the justification for these procedures, however, rests on their accurate application to two simple systems: the H_2^+ ion and the H_2 molecule, the only two molecules to date for which the accuracy of the computations exceeds the precision of the experimental data.

Even for these two molecules, however, the complexity of the mathematics needed for accurate calculations far exceeds the standards required of the reader of this book, and to illustrate the methods used and to give the reader a feel for the numerical magnitudes involved we shall in this chapter explore a synthetic world of one dimension: the delta world.

10.2 THE DELTA FUNCTION

A delta function $\delta(x)$ is an extrapolation of any continuous function of a single variable x which has the property that it has a single, very sharp maximum at the point $x = 0$ and which very rapidly attenuates to zero at all points distant from the origin. We extrapolate this behavior to a "function" $\delta(x)$ which is zero for $x \neq 0$ and which tends to ∞ at $x = 0$ in such a way that

$$\int \delta(x)\, dx = 1$$

in which the range of integration is any range including the origin. It is possible to express the delta function analytically as the limit of a number of recognizable formulas,[1,2] but we shall not need these complications, for all we shall ever require of the delta function is to be able to evaluate integrals in which it appears in the integrand. Thus an integral

$$\int \delta(x) f(x)\, dx$$

in which $f(x)$ is a normal, "well behaved" function of x possesses an integrand which is zero for $x \neq 0$. All of the contribution to the integral therefore arises from the neighborhood of $x = 0$, and we may write formally

$$\int \delta(x) f(x)\, dx = f(0) \int \delta(x)\, dx = f(0) \tag{10.1}$$

The student may thus picture $\delta(x)$ as a single, intense peak located at $x = 0$ which has the virtue that the integral of the product $\delta(x)f(x)$ selects from $f(x)$ the single point $f(0)$.

A more general integral is of the form

$$\int \delta(x - a) f(x) = f(a) \tag{10.2}$$

All that we have done here is to shift the origin of the delta function to the point a, and if the student will glance back to equation 3.3, he will observe that the integral operation

$$\int \delta(y - x) \cdots dy$$

plays the role of the identity operation in function space, for it transforms a function into itself.

10.3 THE DELTA ATOM

I now construct a synthetic world of one dimension in which an electron moves along the x axis in the neighborhood of an atomic nucleus located at the origin. Instead of the usual Coulomb attraction of the electron for the nucleus, I imagine that the attractive potential is proportional to a delta function

$$V(x) = -e^2\delta(x)$$

in which e is the electronic charge. Let us examine the quantum mechanics of this system.

We rapidly set up the Hamiltonian operator

$$H = -\frac{\hbar^2}{2m_e}\frac{d^2}{dx^2} - e^2\delta(x) \tag{10.3}$$

and the Schrödinger equation

$$\frac{d^2\psi}{dx^2} + \frac{2m_e e^2}{\hbar^2}\delta(x)\psi = -\frac{2m_e}{\hbar^2}E\psi \tag{10.4}$$

As was the case for the hydrogen atom, a bound state for the delta atom occurs only if E is negative, and to emphasize this fact let u^2 be the positive quantity $u^2 = -E$. Also letting $2m_e/\hbar^2 = k^2$, equation 10.4 becomes

$$\frac{d^2\psi}{dx^2} + k^2e^2\delta(x)\psi = k^2u^2\psi \tag{10.5}$$

Now for $x \neq 0$, we know from the properties of the delta function that equation 10.5 reduces to

$$\frac{d^2\psi}{dx^2} = k^2u^2\psi$$

whose general solution is

$$\psi = A\exp(kux) + B\exp(-kux)$$

with A and B constants of integration. Because we shall eventually want to normalize ψ, only solutions that vanish at $|x| \to \infty$ are acceptable, which requires that for x negative, $\psi = A\exp(kux)$, and for x positive, $\psi = B\exp(-kux)$. Furthermore, the necessity that the wave function be continuous requires that at $x = 0$ where these two branches meet, $A = B$. Hence

$$\psi = A\exp(-ku|x|)$$

We have as yet made no use of the properties (10.1), (10.2) of the delta function. Let us integrate both sides of equation 10.5 over a small region $-\varepsilon$ to $+\varepsilon$ with ε arbitrarily small:

$$\left.\frac{d\psi}{dx}\right|_{-\varepsilon}^{\varepsilon} + k^2e^2\psi(0) = k^2u^2\int_{-\varepsilon}^{\varepsilon}\psi\,dx \tag{10.6}$$

Letting $\varepsilon \to 0$, the right hand side of equation 10.6 becomes indefinitely small and the left hand side approaches

$$\Delta \frac{d\psi}{dx} + k^2e^2A = 0 \tag{10.7}$$

in which $\Delta(d\psi/dx)$ is the discontinuous change in slope of the two branches of the wave function as they meet at the origin. As $x \to 0$ from the positive side, $d\psi/dx \to -Aku$, and as $x \to 0$ from the negative side, $d\psi/dx \to Aku$. It follows that

$$\Delta \frac{d\psi}{dx} = -2Aku$$

so that from equation 10.7

$$-2ku + k^2e^2 = 0$$

or

$$u = \tfrac{1}{2}ke^2$$

But this means that the energy is

$$E = -u^2 = -\tfrac{1}{4}k^2e^4 = -\frac{1}{2}\frac{m_e e^4}{\hbar^2}$$

which, according to the Bohr formula (5.9), is precisely the ground state energy of the hydrogen atom. Unlike the hydrogen atom, however, the delta atom has no other bound states, and excitation of the electron would result in ionization.

When normalized, the complete wave function for the delta atom is

$$\psi = \left(\frac{k^2e^2}{2}\right)^{1/2} \exp\left(-\tfrac{1}{2}k^2e^2|x|\right) \tag{10.8}$$

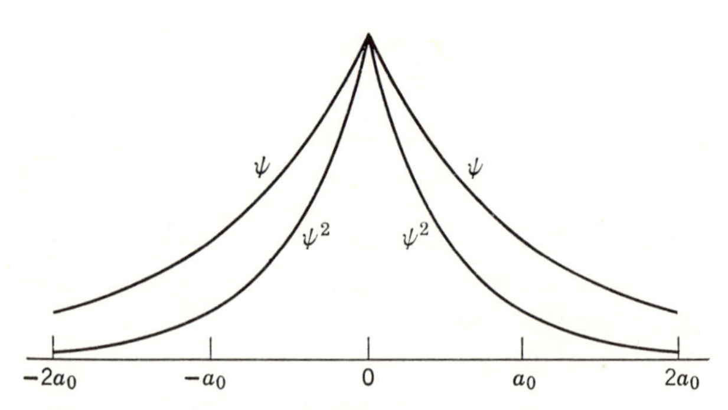

Figure 10.1 Wave function ψ and electron density distribution ψ^2 for the delta atom.

For computational purposes, it is convenient to introduce atomic units of energy and distance. These are

$$a_0 = \frac{2}{k^2e^2} = \frac{\hbar^2}{m_e e^2} = 0.5292\ \text{Å} = \text{the Bohr radius}$$

$$E_0 = \tfrac{1}{2}k^2e^4 = \frac{m_e e^4}{\hbar^2} = \frac{e^2}{a_0} = 627.5\ \text{kcal/mole}$$

In terms of these units the wave function for the delta atom is

$$\psi = a_0^{-1/2} \exp\left(-\frac{|x|}{a_0}\right) \tag{10.9}$$

with energy $E/E_0 = -\frac{1}{2}$. A plot of equation 10.9 is given in Figure 10.1, together with a plot of the electron density distribution ψ^2.

PROBLEM

10.1 Confirm the normalization (10.9) of the delta atom wave function.

10.4 THE DELTA MOLECULE ION

Into the delta world we now introduce some chemistry by considering the motion of a single electron in the vicinity of two delta nuclei located a distance R apart. To maintain some semblance of reality, it will be necessary to let the two nuclei repel each other, and I shall assume that they do this via a damped Coulomb repulsion potential $(e^2/R) \exp(-R/a_0)$. (This choice is frankly governed by the empirical fact that the resulting numerical magnitudes involved in chemical bond formation are roughly comparable to those observed in the real, three dimensional, Coulomb world of the laboratory.) I locate the origin of our coordinate system at a point midway between the two nuclei, and then the Hamiltonian operator for the system is

$$H = -\frac{\hbar^2}{2m_e}\frac{d^2}{dx^2} - e^2\delta(x + \tfrac{1}{2}R) - e^2\delta(x - \tfrac{1}{2}R) + \frac{e^2}{R}\exp\left(-\frac{R}{a_0}\right) \tag{10.10}$$

Note the positive sign in front of the internuclear repulsion potential.

For the Schrödinger equation $H\psi = E\psi$ we have

$$\frac{d^2\psi}{dx^2} + k^2e^2[\delta(x + \tfrac{1}{2}R) + \delta(x - \tfrac{1}{2}R)]\psi = k^2w^2\psi$$

in which $k^2 = 2m_e/\hbar^2$ and

$$w^2 = -E + \frac{e^2}{R}\exp\left(-\frac{R}{a_0}\right) \tag{10.11}$$

As was the case for the delta atom, the delta functions vanish at all points $x \neq \pm\frac{1}{2}R$, and this means that the wave function is built up out of pieces

$$\begin{aligned} \psi &= A \exp(kwx) \quad \text{for} \quad x < -\tfrac{1}{2}R \\ \psi &= B \exp(kwx) + C \exp(-kwx) \quad \text{for} \quad -\tfrac{1}{2}R \leq x \leq \tfrac{1}{2}R \\ \psi &= D \exp(-kwx) \quad \text{for} \quad x > \tfrac{1}{2}R \end{aligned} \tag{10.12}$$

for no other choice will result in a normalizable function. To fit these pieces together, we apply conditions of continuity of the wave function at $x = \pm\frac{1}{2}R$. By integrating the Schrödinger equation over an ε region in the neighborhood of each nucleus and making use of (10.2), we also find that the pieces must be joined together with discontinuous slope

$$\Delta \frac{d\psi}{dx} = -k^2 e^2 \psi \tag{10.13}$$

at $x = \pm\frac{1}{2}R$. These conditions lead to a set of four linear, homogeneous equations in the integration constants A, B, C, D, for which a nontrivial solution exists only if the determinant of coefficients vanishes. Upon expansion of the 4×4 determinant, it turns out that w must be chosen to satisfy

$$w = \tfrac{1}{2}ke^2[1 \pm \exp(-kwR)] \tag{10.14}$$

Equation 10.14 is transcendental in w with an apparent choice of two roots depending upon which sign is chosen. For the negative sign, the root is always $w = 0$ no matter what the separation R of the nuclei, and this in turn from (10.11) means that

$$E_- = \frac{e^2}{R} \exp\left(-\frac{R}{a_0}\right)$$

which is to say that the total energy of the system is always positive, leading to no molecule formation. E_- is thus an antibonding state, for the nuclei would tend to fly apart.

The choice of the positive sign in equation 10.14 is more interesting. We note first of all that if $R \to \infty$, then $w \to \frac{1}{2}ke^2$, that is

$$E_+ \to -w^2 = -\tfrac{1}{4}k^2e^4 = -\tfrac{1}{2}E_0$$

This is precisely the energy of an isolated delta atom, a result fully in accord with the chemist's intuitive understanding of chemical bond formation. For finite R, we choose w to satisfy

$$w = \tfrac{1}{2}ke^2[1 + \exp(-kwR)]$$

Table 10.1 Exact Calculations for the Delta Molecule Ion

ρ	y	y^2	$(1/\rho)\exp(-\rho)$	E_+/E_0
0.0	1.4142	2.0000	∞	∞
0.5	1.0449	1.0918	1.2131	+0.1213
1.0	0.9040	0.8172	0.3679	−0.4493
1.2	0.8690	0.7552	0.2510	−0.5042
1.4	0.8409	0.7071	0.1761	−0.5310
1.6	0.8181	0.6693	0.1262	−0.5431
1.7	0.8084	0.6535	0.1075	−0.5460
1.8	0.7995	0.6392	0.0918	−0.5474
1.9	0.7914	0.6263	0.0787	−0.5476
2.0	0.7841	0.6148	0.0677	−0.5471
2.2	0.7712	0.5947	0.0504	−0.5443
2.4	0.7606	0.5785	0.0378	−0.5407
3.0	0.7380	0.5446	0.0166	−0.5280
∞	0.7071	0.5000	0.0000	−0.5000

In atomic units $y = w/\sqrt{E_0} = \sqrt{2}w/ke^2$ and $\rho = R/a_0 = \frac{1}{2}k^2e^2R$, this equation becomes

$$y = \frac{1}{\sqrt{2}}[1 + \exp(-\sqrt{2}\,\rho y)] \tag{10.15}$$

and a selection of roots y of equation 10.15 as a function of ρ is given in Table 10.1. Also tabulated in atomic units are the internuclear repulsion potential $(1/\rho)\exp(-\rho)$ and the total energy $E_+/E_0 = (1/\rho)\exp(-\rho) - y^2$ calculated from equation 10.11.

PROBLEM

10.2 Construct from the pieces (10.12) of the wave function of the delta molecule ion a set of four linear, homogeneous equations in A, B, C, D which satisfy the condition that the wave function be continuous at $x = \pm\frac{1}{2}R$ and that the slopes of the wave function at these points be discontinuous by an amount (10.13). Set the determinant of coefficients of these equations equal to zero and recover the formula (10.14).

10.5 THE POTENTIAL CURVE

A very interesting result is obtained if the total energy E_+/E_0 is plotted against the internuclear distance $\rho = R/a_0$. Employing data from Table 10.1, Figure 10.2 is obtained, whence it will be observed that the system

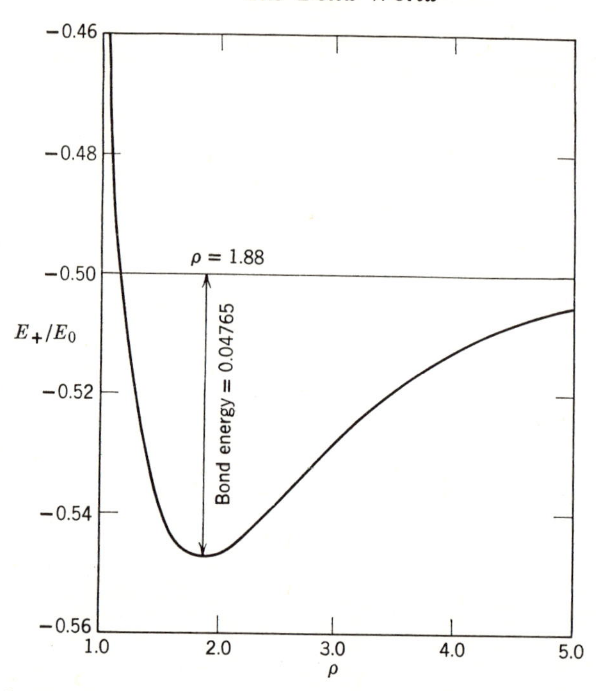

Figure 10.2 Potential energy curve for the delta molecule ion. E_+/E_0 and ρ are in atomic units.

consisting of two delta nuclei plus one electron achieves a minimum energy at $\rho = 1.880$ ($R = 1.880 \times 0.5292 = 0.9949$ Å) and that the total energy at the minimum is $E_+/E_0 = -0.54765$. Because this is lower by 0.04765 atomic units than the energy of the infinitely separated nuclei, we may state that the bond energy of the delta molecule ion is $0.04765 \times 627.5 = 29.90$ kcal/mole.

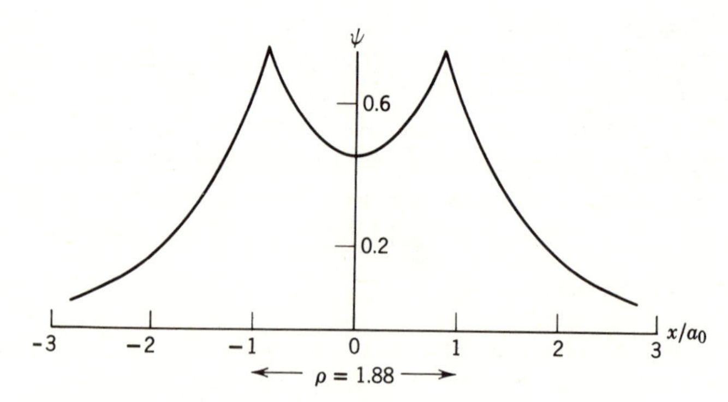

Figure 10.3 Bonding wave function for the delta molecule ion.

Potential curves of the sort sketched in Figure 10.2 are also the result of the more elaborate calculations which have been performed for simple, diatomic molecules in the Coulomb world of the laboratory. They inevitably accompany wave functions which indicate a build up of electron density between the bonded nuclei, and such wave functions are in general called bonding molecular orbitals. As an example, Figure 10.3 shows the normalized wave function for the delta molecule ion at the equilibrium internuclear separation $\rho = 1.880$. For those systems H_2^+ and H_2 for which the corresponding calculations are available, the major contribution to the energy of the chemical bond is found to arise from the fact that the bonding molecular orbital permits the electrons to occupy in high density the region of low potential energy between the nuclei.

PROBLEM

10.3 The last statement may be checked for the delta world by calculating separately the expectation of the potential and of the kinetic energies. Thus for the delta atom

$$V = -e^2\delta(x)$$

and

$$\langle V \rangle = \int V(x)\psi^2(x)\, dx$$

$$= -\frac{e^2}{a_0}\int \delta(x) \exp\left(-2\frac{|x|}{a_0}\right) dx$$

$$= -\frac{e^2}{a_0} = -E_0$$

But the total energy of the atom is $E = \langle K \rangle + \langle V \rangle = -\frac{1}{2}E_0$, whence for the expectation of the kinetic energy we have

$$\langle K \rangle = E - \langle V \rangle = \tfrac{1}{2}E_0$$

Given the normalized wave function for the delta molecule ion at the equilibrium bond distance $R = 1.880a_0$,

$$\sqrt{a_0}\,\psi = (0.7155)\exp\left[1.1214\left(\frac{x}{a_0} + 0.940\right)\right]; \qquad \frac{x}{a_0} \leq -0.940$$

$$\sqrt{a_0}\,\psi = (0.4447)\cosh\left(1.1214\frac{x}{a_0}\right); \qquad -0.940 \leq \frac{x}{a_0} \leq 0.940$$

$$\sqrt{a_0}\,\psi = (0.7155)\exp\left[-1.1214\left(\frac{x}{a_0} - 0.940\right)\right]; \qquad \frac{x}{a_0} \geq 0.940$$

calculate the expectation of V and of K in the molecule ion.

Your answers should be $\langle V \rangle = -0.9427E_0$ and $\langle K \rangle = 0.3951E_0$. Note that $\langle V \rangle$ is thus actually higher in the molecule ion than in the atom, hence that in the delta world the energy of a chemical bond derives exclusively from the lowering of the kinetic energy of the electron when it is delocalized from the vicinity of a single nucleus into the larger space in the vicinity of two adjacent nuclei. This effect also exists in the Coulomb world, but it contributes less to the total bond energy than the purely electrostatic one described in the last sentence of Section 10.5.

10.6 LCAO

Because accurate calculations for real molecules are immensely complex, we turn to the variation method in the search for approximate results. Even here the computational demands on the investigator are formidable, and generally speaking in the chapters to follow we shall not even complete what is formally demanded by the variation method. However, because the qualitative arguments used in molecular theory rest on the experience gained in their accurate application to a few simple systems, we shall in the remainder of this chapter carry out a complete LCAO analysis of the delta molecule ion.

LCAO stands for linear combination of atomic orbitals, meaning that we choose as a variation form (9.4) for our molecular orbital a superposition of atomic orbitals centered on the different nuclei. This form is suggested by a comparison of Figures 10.1 and 10.3, for if an atomic wave function centered on nucleus a is added to the identical function centered on b, the sum of the two bears a strong qualitative resemblance to the exact molecular orbital (see Figure 10.4). Explicitly this means

$$\psi = c_1\phi_a + c_2\phi_b$$

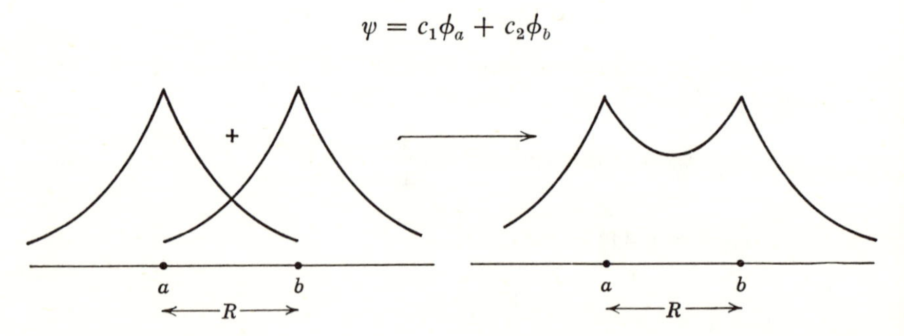

Figure 10.4 Synthesis of an LCAO variation wave function from atomic orbitals centered on two identical nuclei.

with

$$\phi_a = a_0^{-1/2} \exp\left(-\frac{|x + \frac{1}{2}R|}{a_0}\right)$$

$$\phi_b = a_0^{-1/2} \exp\left(-\frac{|x - \frac{1}{2}R|}{a_0}\right)$$

and the c's are to be calculated via the variation program of Section 9.9.

The required matrix elements are

$$H_{11} = \int \phi_a H \phi_a \, dx \equiv \alpha$$

$$H_{22} = \int \phi_b H \phi_b \, dx = \alpha$$

$$H_{12} = H_{21} = \int \phi_a H \phi_b \, dx \equiv \beta$$

$$S_{11} = \int \phi_a^2 \, dx = 1$$

$$S_{22} = \int \phi_b^2 \, dx = 1$$

$$S_{12} = S_{21} = \int \phi_a \phi_b \, dx \equiv S$$

in which H is the molecular Hamiltonian (10.10). Equality of H_{11} and H_{22} follows from the fact that each of the nuclei is identical, and $S_{11} = S_{22} = 1$ because the basis functions ϕ_a and ϕ_b are normalized. They are not, however, orthogonal, and this fact gives rise to the so-called *overlap integral* S. The energy levels are the roots of the secular equation

$$\begin{vmatrix} \alpha - E & \beta - ES \\ \beta - ES & \alpha - E \end{vmatrix} = 0 \qquad (10.16)$$

which is to say

$$E_1 = \frac{\alpha + \beta}{1 + S}; \qquad E_2 = \frac{\alpha - \beta}{1 - S} \qquad (10.17)$$

The student will also find without difficulty that the appropriate normalized molecular orbitals for these energy levels are

$$\psi_1 = \frac{1}{\sqrt{2 + 2S}}(\phi_a + \phi_b); \qquad \psi_2 = \frac{1}{\sqrt{2 - 2S}}(\phi_a - \phi_b) \qquad (10.18)$$

10.7 CALCULATION OF THE INTEGRALS

Further progress requires explicit calculation of the matrix elements as functions of R. I take up first the integral α, called a *Coulomb integral*,* for it involves only one of the two atomic wave functions:

$$\alpha = H_{11} = \int \phi_a H \phi_a \, dx$$

$$= \int_{-\infty}^{\infty} \phi_a \left\{ -\frac{1}{k^2}\frac{d^2}{dx^2} - e^2\delta(x + \tfrac{1}{2}R) - e^2\delta(x - \tfrac{1}{2}R) + \frac{e^2}{R}\exp\left(-\frac{R}{a_0}\right) \right\} \phi_a \, dx$$

$$= \int_{-\infty}^{\infty} \phi_a \left\{ -\frac{1}{k^2}\frac{d^2}{dx^2} - e^2\delta(x + \tfrac{1}{2}R) \right\} \phi_a \, dx$$

$$- e^2 \int_{-\infty}^{\infty} \phi_a^2 \delta(x - \tfrac{1}{2}R)\, dx + \frac{e^2}{R}\exp\left(-\frac{R}{a_0}\right) \int_{-\infty}^{\infty} \phi_a^2 \, dx \qquad (10.19)$$

Of the three integrals in the last line of equation 10.19, the first is the expectation $-\frac{1}{2}E_0$ of the atomic Hamiltonian (10.3) for a delta atom located at $x = -\frac{1}{2}R$ and the last integral has already been normalized to unity:

$$\alpha = \int \phi_a H \phi_a \, dx = -\tfrac{1}{2}E_0 - e^2\phi_a^2(\tfrac{1}{2}R) + \frac{e^2}{R}\exp\left(-\frac{R}{a_0}\right)$$

$$= -\tfrac{1}{2}E_0 - \frac{e^2}{a_0}\exp\left(-2\frac{R}{a_0}\right) + \frac{e^2}{R}\exp\left(-\frac{R}{a_0}\right)$$

Switching to atomic units this becomes

$$\frac{\alpha}{E_0} = -\tfrac{1}{2} - \exp(-2\rho) + \frac{1}{\rho}\exp(-\rho)$$

The overlap integral is a little trickier:

$$S = \int \phi_a \phi_b \, dx = a_0^{-1} \int_{-\infty}^{\infty} \exp\left(-\frac{|x + \frac{1}{2}R|}{a_0}\right) \exp\left(-\frac{|x - \frac{1}{2}R|}{a_0}\right) dx$$

Break up the range of integration into three regions;

$$\int_{-\infty}^{\infty} = \int_{-\infty}^{-R/2} + \int_{-R/2}^{R/2} + \int_{R/2}^{\infty}$$

and substitute into each the appropriate form of the integrand. After some

* The Coulomb integral α which arises in LCAO theory is not the same as that defined in equation 8.6. In LCAO theory α requires an integration over the configuration space of a single electron. In (8.6) J is the result of an integration over the configuration space of two electrons.

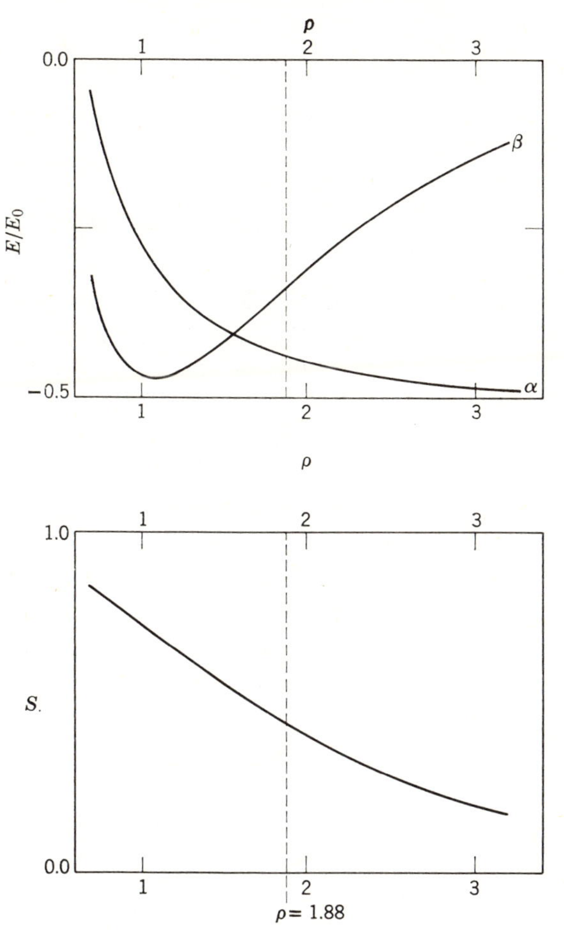

Figure 10.5 Coulomb α, resonance β, and overlap integrals S as functions of the internuclear separation ρ in the delta molecule ion.

simplification, the result in atomic units is

$$S = (1 + \rho) \exp(-\rho) \tag{10.20}$$

The *resonance integral* β yields to the same method used for α:

$$\begin{aligned} \beta &= \int \phi_a H \phi_b \, dx \\ \frac{\beta}{E_0} &= -\tfrac{1}{2}(3 + \rho) \exp(-\rho) + \left(1 + \frac{1}{\rho}\right) \exp(-2\rho) \end{aligned} \tag{10.21}$$

Plots of these integrals as functions of the internuclear separation ρ are drawn in Figure 10.5.

PROBLEM

10.4 Derive formulas (10.20) and (10.21) for the overlap and resonance integrals.

10.8 BONDING AND ANTIBONDING MOLECULAR ORBITALS

From the formulas of Section 10.7 we can calculate the approximate energy levels (10.17) and their associated wave functions (10.18). E_1 and E_2 are plotted in atomic units in Figure 10.6 together with a portion of the exact potential curve of Figure 10.2. In accordance with the variation theorem, the LCAO result lies above the exact curve and predicts a bond distance $\rho = 2.01$ ($R = 1.063$ Å) with associated molecular energy $E_1/E_0 = -0.54162$ ($E_1 = -339.86$ kcal/mole). These figures are compared with the

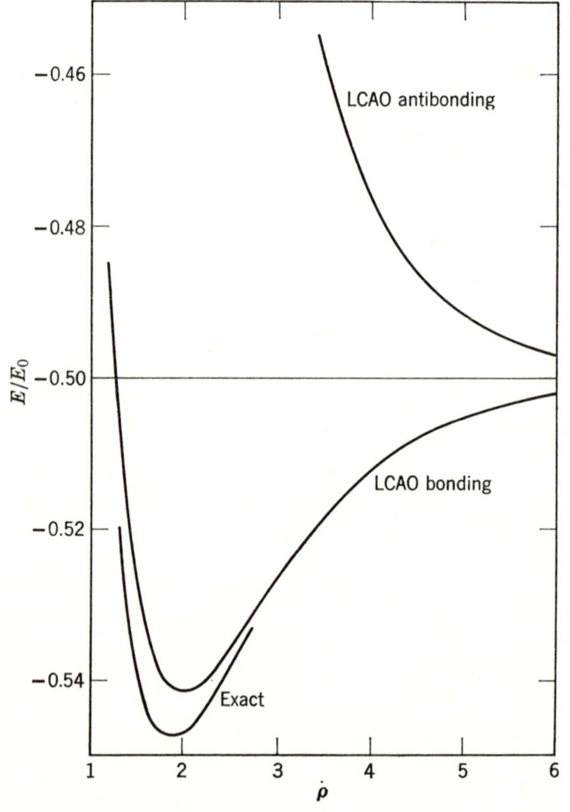

Figure 10.6 A comparison of the LCAO approximate potential curve with the exact result for the delta molecule ion.

Table 10.2

	Bond distance (Å)	Molecular energy (kcal/mole)	Bond energy (kcal/mole)
Exact	0.9949	−343.65	29.90
LCAO	1.0634	−339.86	26.11
Percent error	7%	1%	13%

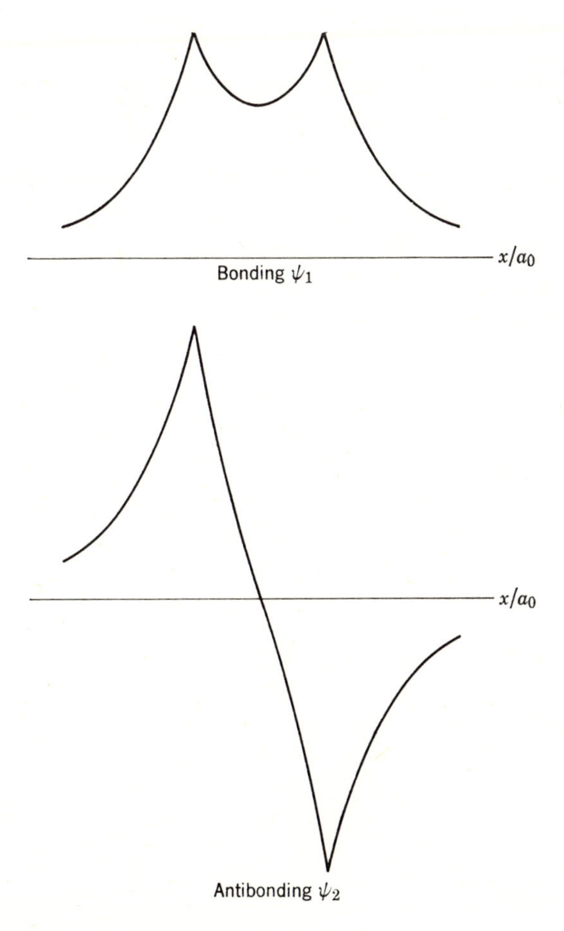

Figure 10.7 Bonding and antibonding LCAO molecular orbitals for the delta molecule ion.

exact results in Table 10.2, whence it will be observed that while we have accounted for the major features of chemical bonding by using the LCAO method, we are a long way from being quantitative. The error in the total energy is not so bad, but because the bond energy is only a small fraction of the total molecular energy, a small error in the first can be magnified into a major one in the second. The results are, however, qualitatively good enough so that we may with confidence call ψ_1 a bonding LCAO molecular orbital.

Because the root E_2 always lies above the energy of a solitary delta atom, it represents repulsion at all distances between the two delta nuclei and thus cannot lead to chemical bond formation. It is therefore appropriate to call ψ_2 an antibonding LCAO molecular orbital.

The qualitative difference between bonding and antibonding molecular orbitals is strikingly illustrated by the plots of Figure 10.7. The LCAO bonding orbital ψ_1 is readily identifiable from its resemblance to the exact profile of Figure 10.3. It is the minus sign for ψ_2 in equation 10.18 which has led to the distinguishing feature of an antibonding molecular orbital: the presence of a node cutting the internuclear axis between the nuclei. When electron density profiles are calculated by squaring the wave function, this node will persist in the form of a region of zero electron density severing the bond axis; and it follows that if in the Coulomb world ψ_1 is bonding because it places a high concentration of charge in the region of low potential energy between the nuclei, then ψ_2 is antibonding because it forces charge away from this vital domain.

In Figure 10.8 are sketched the electron density profiles ψ^2 for the exact and LCAO bonding wave functions. They agree reasonably well, but the reason that the LCAO energy E_1 always lies above the exact molecular energy is directly ascribable to the fact that the LCAO approximation predicts a slightly lower electron density between the nuclei than is actually the case.

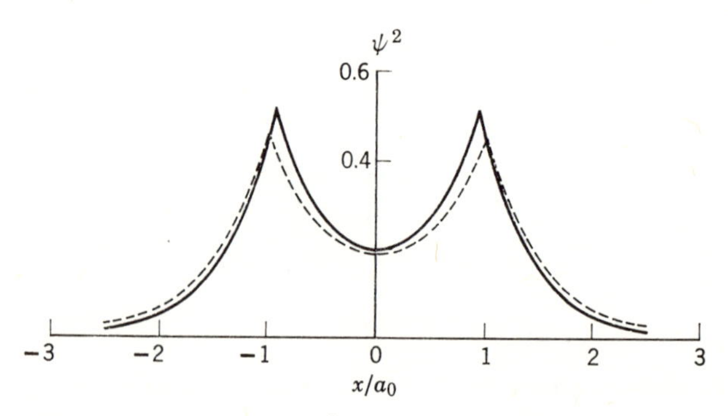

Figure 10.8 Electron density profiles ψ^2, exact (——) and LCAO (- - -), for the delta molecule ion.

10.9 INTERPRETATION OF THE INTEGRALS

Let us now turn back to Figure 10.5 and search for a handy interpretation of the Coulomb and resonance integrals. At $R = \infty$, the resonance integral β is zero and the Coulomb integral α is the atomic energy $\alpha = -\frac{1}{2}E_0$. Even at the equilibrium bond distance, α is only 10% away from the original atomic energy, meaning that even in the molecule, most of the contribution to α comes from that part of the molecular Hamiltonian which can be identified as the Hamiltonian of a single atom. In classifying these integrals it is therefore customary to identify approximately the Coulomb integrals with the energy of an electron in the parent atom before molecule formation takes place.

There is no atomic interpretation of the resonance integral β. In the separated atoms β is zero; and at equilibrium bond distance, β is significantly large, of the order of magnitude of α. The resonance integral is therefore a distinctively molecular quantity and is responsible for the primary contribution to the bond energy. This fact becomes particularly clear if we make the extremely crude approximation of neglecting the overlap integral S in equations 10.17. The molecular energies become

$$E_1 = \alpha + \beta; \qquad E_2 = \alpha - \beta \tag{10.22}$$

meaning that the degenerate atomic energy levels α, α of an electron on either of two nuclei at infinite separation split into a bonding and an antibonding level when the two nuclei are brought close to each other (Figure 10.9). Although we have as yet made no formal contact with the Hückel molecular orbital theory developed in Chapter 2, the reader cannot help but notice the similarity of these results to those obtained for the ethylene molecule of Section 2.4, Figure 2.2.

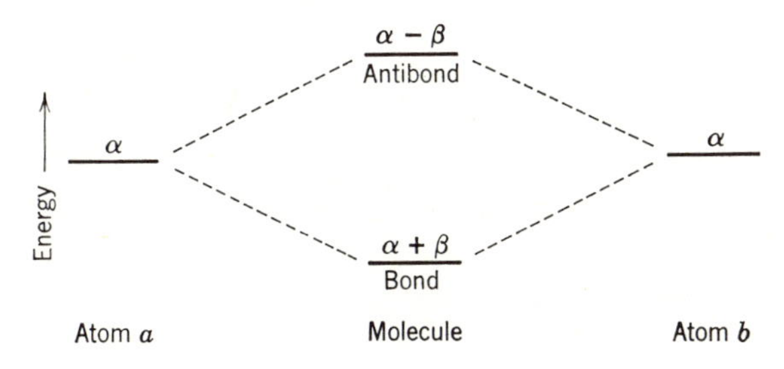

Figure 10.9 Schematic energy level diagram in the delta molecule ion, LCAO approximation.

Neglect of overlap, which amounts to the assumption of orthogonality between ϕ_a and ϕ_b at all distances R, is a very poor quantitative approximation, but at least it does not destroy the qualitative ordering of bonding and antibonding levels in the molecule, and we shall make increasing use of this approximation in qualitative work. It has the great computational advantage of simplifying secular equations of the form (9.5) into those of the form (9.7).

10.10 BOND ENERGY AND OVERLAP

An interesting property of the overlap integral S is revealed by plotting it parametrically against the bond energy (Figure 10.10). As the nuclei are brought together from infinity, bond energy and overlap are seen to be

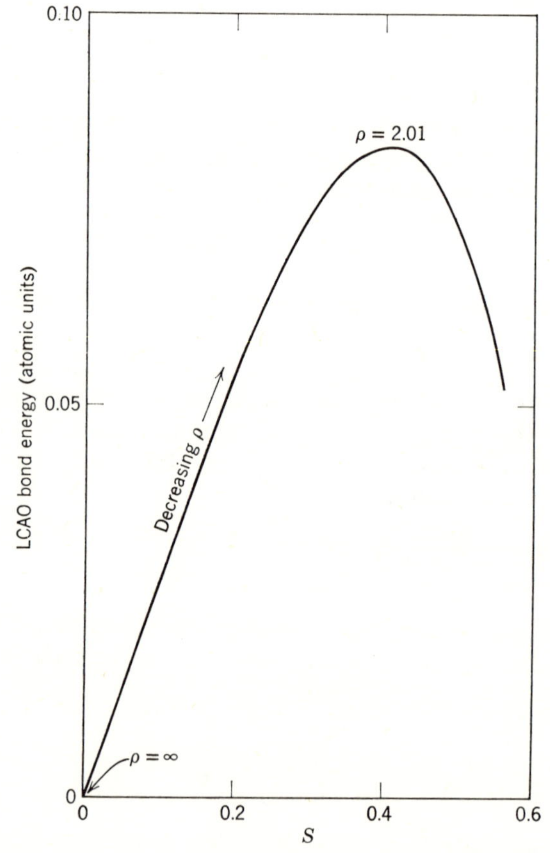

Figure 10.10 Parametric plot of the bond energy as a function of the overlap integral S.

accurately proportional until quite close to the equilibrium bond distance. This empirical observation has guided the choice of atomic orbitals for use in the LCAO approximation through the criterion that the strongest bonds are formed from those atomic orbitals which overlap the most when the nuclei are positioned at their equilibrium distances. We shall have occasion to observe this rule at work when we come to the construction of hybrid atomic orbitals for use in valence theory.

REFERENCES

1. P. M. Morse and H. Feshbach, *Methods of Theoretical Physics*, McGraw-Hill, Inc., New York, 1953, p. 122.
2. I. N. Sneddon, *Fourier Transforms*, McGraw-Hill, Inc., New York, 1951, pp. 32–34.

Chapter 11

DIATOMIC MOLECULES

11.1 THE H_2^+ ION

The hydrogen molecule ion is the traditional jumping off place for the beginning student of molecular quantum mechanics. Because we have already lavished a good deal of attention on its delta analog in Chapter 10, there will be no quantitative exploration of H_2^+ here. Instead I shall rely on the student's ability to generalize imaginatively from his experience in the delta world.

Let two protons a, b be placed a distance R apart on the z axis of a Cartesian coordinate system and upon this framework cast a single electron. The coordinate diagram of Figure 11.1 and the Hamiltonian (11.1) should be self explanatory.

$$H = -\frac{\hbar^2}{2m_e}\nabla^2 - \frac{e^2}{r_a} - \frac{e^2}{r_b} + \frac{e^2}{R} \tag{11.1}$$

If the Cartesian coordinates are transformed into confocal, elliptical coordinates[1–3] $(r_a + r_b)/R$, $(r_a - r_b)/R$, φ, the Schrödinger equation proves to be separable for a trial wave function of the form

$$\psi = \exp(\pm im\varphi)\Phi(r_a, r_b, R); \qquad m = 0, 1, \ldots \tag{11.2}$$

in which φ is an angle running from 0 to 2π measured around the internuclear axis. The reason for this separation is that the potential energy of the system does not depend on φ, and the appearance of an integral quantum number m follows from the familiar requirement that ψ be continuous as φ increases through 2π.

Knowing only (11.2) we are already able to anticipate something about the form of the molecular wave functions by classifying them as $\sigma, \pi, \delta, \ldots$ molecular orbitals according to whether $|m| = 0, 1, 2, \ldots$. The energy

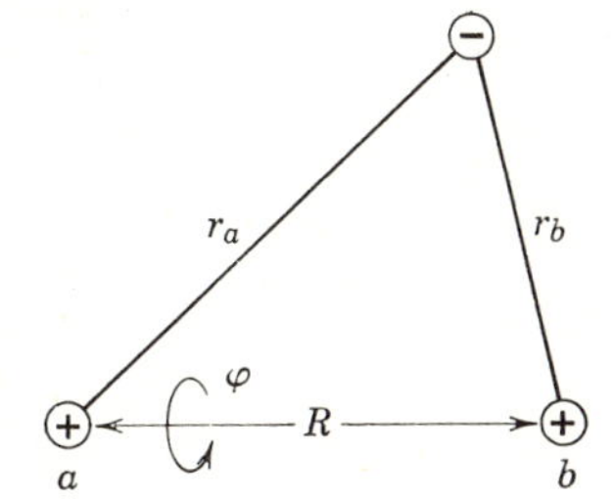

Figure 11.1 The H_2^+ ion.

turns out to be independent of the sign of m, so that while σ states are non-degenerate, each π, δ, and so on, state is doubly degenerate. The molecular wave functions are easier to picture if we take appropriate linear combinations of the complex exponentials exp $(im\varphi)$ and exp $(-im\varphi)$ for each degenerate state. As was the case for the isotropic oscillator of Chapter 4, the resulting wave functions are real, orthogonal, and have the pictorial convenience that they exhibit $|m|$ nodal planes passing through the internuclear axis. Looking at the molecule end on by putting the bond axis perpendicular to the plane of the paper, the nodal patterns of σ, π, and δ states are sketched in Figure 11.2. These nodal diagrams are to be compared with Figures 2.3, 4.4, and 4.5, derived for benzene and for the isotropic oscillator. They are characteristic of wave functions for which an electron moves in a field of potential energy which has an axis of symmetry. At the same time, the student must remember the arbitrariness of the assignment of nodal planes, for while we can say with confidence that a π state possesses only a single nodal plane, the degeneracy of this level in a diatomic molecule prevents us from knowing anything about how the plane is oriented.

We can tell something else about the way the wave functions will look by examining the symmetry of the Hamiltonian operator upon reflection of the coordinate system through a plane perpendicular to the internuclear axis at the midway point. For z the bond axis and $z = 0$ halfway between the nuclei, the quantities r_a and r_b defined by Figure 11.1 are

$$\begin{aligned} r_a &= [x^2 + y^2 + (z + \tfrac{1}{2}R)^2]^{1/2} \\ r_b &= [x^2 + y^2 + (z - \tfrac{1}{2}R)^2]^{1/2} \end{aligned} \qquad (11.3)$$

If in these equations z is replaced by $-z$, we have effectively reflected our coordinate system through the xy coordinate plane, and the algebraic effect in equations 11.3 is to change r_a into r_b. Such a transformation has no effect upon the Hamiltonian operator (11.1), for it only permutes the order of the terms as they appear in the potential energy. In a similar way, the operator

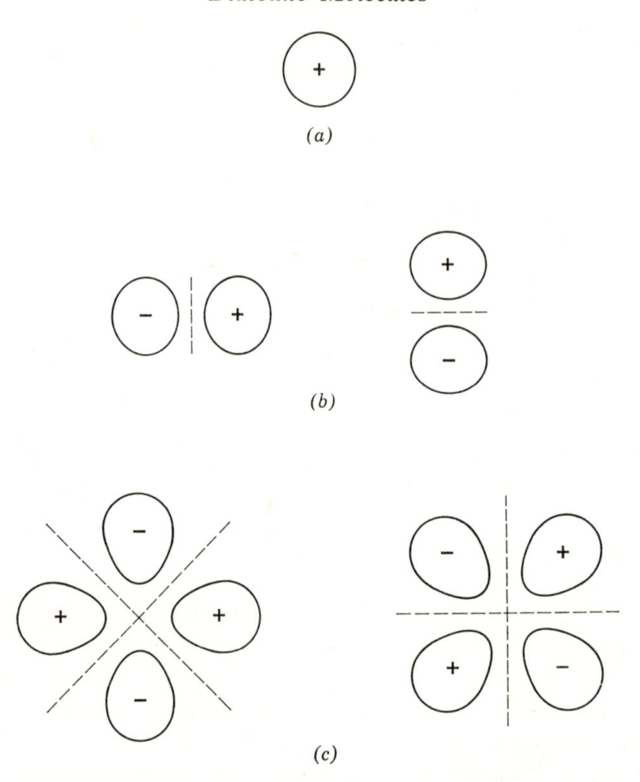

Figure 11.2 Molecular states showing nodal planes. The bond axis is perpendicular to the plane of the paper. (*a*) σ state (nondegenerate); (*b*) π states (doubly degenerate); (*c*) δ states (doubly degenerate).

$\partial^2/\partial z^2$ which occurs in the Laplacian operator is unaffected by changing the sign of z. It follows that the Hamiltonian operator of the hydrogen molecule ion is invariant upon reflection through the xy plane.

Let us now consider the Schrödinger equation for a nondegenerate, σ state of the H_2^+ ion:

$$H(x, y, z)\psi(x, y, z) = E\psi(x, y, z)$$

If we change the sign of z on each side of the equation, it becomes

$$H(x, y, z)\psi(x, y, -z) = E\psi(x, y, -z)$$

H, as we have just seen, is unaffected by the reflection operation. But because the state is nondegenerate, its eigenfunction is to within a multiplicative constant unique, and this can only mean that $\psi(x, y, -z)$ is a multiple of $\psi(x, y, z)$. If normalization is to be maintained, the only possible multiples are $+1$ or -1, that is, either

$$\psi(x, y, z) = \psi(x, y, -z)$$

or

$$\psi(x, y, z) = -\psi(x, y, -z) \tag{11.4}$$

In the first case the wave function is said to be symmetric with respect to reflection through the xy plane, and in the second case the wave function is said to be antisymmetric. The latter functions have the interesting property that they are zero everywhere in the xy plane, as can easily be proven by setting $z = 0$ in equation 11.4.

While I have demonstrated this symmetry property only for the nondegenerate, σ states, it proves to be valid also for the degenerate π and δ states; and a complete solution[4–7] of the Schrödinger equation for H_2^+ reveals the existence of both types of symmetry at all energy levels. We shall designate the antisymmetric type by an asterisk. Thus a σ molecular orbital is without nodes passing through or perpendicular to the bond axis in the internuclear region. A σ^* molecular orbital is symmetric around the bond axis and possesses a node perpendicular to the axis between the two nuclei. A π molecular orbital has a single node which includes the bond axis. A π^* molecular orbital adds to this a node severing the axis.

The alert student should by now have been reminded of the LCAO treatment of the delta molecule ion (Section 10.8). Figure 10.7 already shows the reflection symmetry and antisymmetry which we have here predicted for the H_2^+ ion, and we are immediately led to suspect the chemical significance of starred and unstarred molecular orbitals, with the latter antibonding in character and the former bonding. The analysis of Section 11.2 will confirm this generalization.

11.2 LCAO FOR THE H_2^+ ION

A knowledge of the symmetry types of wave function predicted for an exact solution of the wave equation for H_2^+ is an invaluable guide in constructing approximate wave functions via the LCAO technique. Suppose that we wish to examine the ground state of the molecule. All of our experience to date has suggested that the greater the number of nodes in a wave function, the higher will be its energy, so that the ground state should be of the σ classification. A linear combination

$$\psi = c_1(1s)_a + c_2(1s)_b \tag{11.5}$$

of two $1s$ hydrogen atomic orbitals centered on the nuclei a and b has the proper symmetry, being a superposition on two different centers of two spherically symmetric charge distributions (Figure 11.3).

The variation wave function (11.5) leads to a secular equation identical in form with (10.16),

$$\begin{vmatrix} \alpha - E & \beta - ES \\ \beta - ES & \alpha - E \end{vmatrix} = 0$$

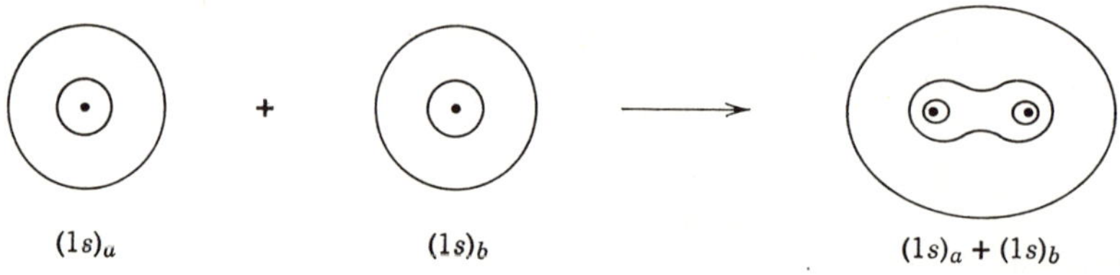

Figure 11.3 Linear superposition of two hydrogen $1s$ atomic orbitals in a bonding molecular orbital.

where

$$\begin{aligned} \alpha &= \int (1s)_a H (1s)_a \, d\tau \qquad \text{is a Coulomb integral} \\ \beta &= \int (1s)_a H (1s)_b \, d\tau \qquad \text{is a resonance integral} \\ S &= \int (1s)_a (1s)_b \, d\tau \qquad \text{is an overlap integral} \end{aligned} \tag{11.6}$$

The Hamiltonian H is given by (11.1), and for completeness the atomic orbitals are explicitly

$$\begin{aligned} (1s)_a &= \pi^{-1/2} a_0^{-3/2} \exp\left(-\frac{r_a}{a_0}\right) \\ (1s)_b &= \pi^{-1/2} a_0^{-3/2} \exp\left(-\frac{r_b}{a_0}\right) \end{aligned} \tag{11.7}$$

We are therefore in the same position with regard to LCAO calculations for H_2^+ as we were at the end of Section 10.6 for the delta molecule ion. All that needs to be done is to calculate the integrals (11.6), and then the approximate energy levels and wave functions will be

$$\begin{aligned} E_\sigma &= \frac{\alpha + \beta}{1 + S}; \qquad \sigma 1s = \frac{1}{\sqrt{2 + 2S}} [(1s)_a + (1s)_b] \\ E_\sigma^* &= \frac{\alpha - \beta}{1 - S}; \qquad \sigma^* 1s = \frac{1}{\sqrt{2 - 2S}} [(1s)_a - (1s)_b] \end{aligned} \tag{11.8}$$

The splitting of the originally doubly degenerate atomic energy levels into a lower bonding level σ and an antibonding level σ^* follows exactly the same pattern as that sketched in Figure 10.9 for the delta molecule ion, and to accompany Figure 11.3 for $\sigma 1s$, we have Figure 11.4 for the antibonding orbital $\sigma^* 1s$.

By casually stating in the preceding paragraph that "all that needs to be done is to calculate the integrals . . . ," however, we have symbolically

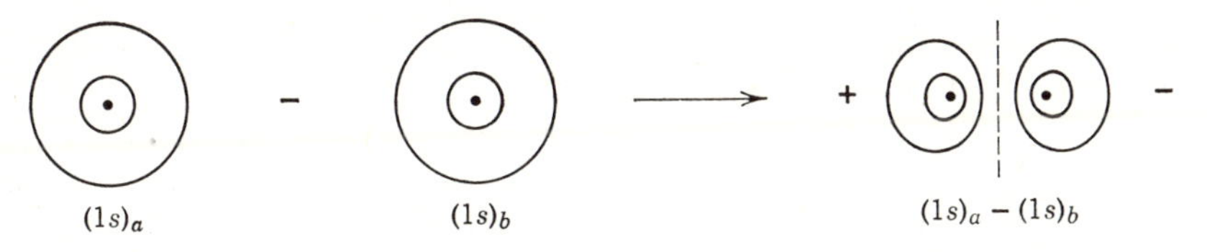

Figure 11.4 Superposition of two hydrogen 1s atomic orbitals in an antibonding molecular orbital.

washed our hands of a very difficult task. For H_2^+ the required integrals have been worked out[7–9] as explicit functions of R, and calculations of this sort for many-electron molecules have engaged the attention of some of the ablest theoretical chemists over the past forty years. In this book we shall take slight interest in these difficult details, preferring to rest our case upon a qualitative interpretation of the quantities α, β, S without ever calculating them explicitly. These are that the Coulomb integral α is approximately the energy of an electron in an atomic orbital before molecule formation takes place, that the resonance integral β is zero at infinite separation of the nuclei and becomes negative and comparable in size with α near the equilibrium bond distance, that the bond energy of a bonding molecular orbital is roughly proportional to the overlap S, so that the strongest bonds and antibonds should arise from an LCAO wave function whose constituent atomic orbitals overlap the most.

11.3 EXCITED STATES OF H_2^+

For the delta molecule ion there was little variety that could be introduced into the LCAO method, for the delta atom possesses only a single bound state. The hydrogen atom, however, possesses an infinity of such states, and this opens the way to more exotic possibilities. Consider, for example, a linear combination of 2s atomic orbitals centered on each nucleus,

$$\psi = c_1(2s)_a + c_2(2s)_b$$

Our experience by now is such that we may immediately infer approximate energy levels and wave functions

$$E_\sigma = \frac{\alpha + \beta}{1 + S}; \qquad \sigma 2s = \frac{1}{\sqrt{2 + 2S}}[(2s)_a + (2s)_b]$$
$$E_\sigma^* = \frac{\alpha - \beta}{1 - S}; \qquad \sigma^* 2s = \frac{1}{\sqrt{2 - 2S}}[(2s)_a - (2s)_b] \qquad (11.9)$$

together with the wave function contours of Figure 11.5. The notation of equations 11.9 is identical with that of equations 11.8, but the student should not be deceived into imagining an identity of the numerical results, for the Coulomb, resonance, and overlap integrals for the $\sigma 2s$ and $\sigma^* 2s$ functions are in equations 11.9 to be calculated from the atomic $2s$ functions rather that from the $1s$ functions of equations 11.7. On the face of it, the patterns of Figure 11.5 would not appear to be very different from those drawn in Figures 11.3 and 11.4 for the $\sigma 1s$ and $\sigma^* 1s$ functions, but the student should remember that the superposed atomic $2s$ functions are bigger than the $1s$, meaning the electron distribution lies further away from the nuclei, and also that the atomic $2s$ function possesses a spherical node. This node persists in a complicated, axially symmetric way into the molecular orbitals, and it is drawn as a dotted contour in the $\sigma 2s$ diagram of Figure 11.5.

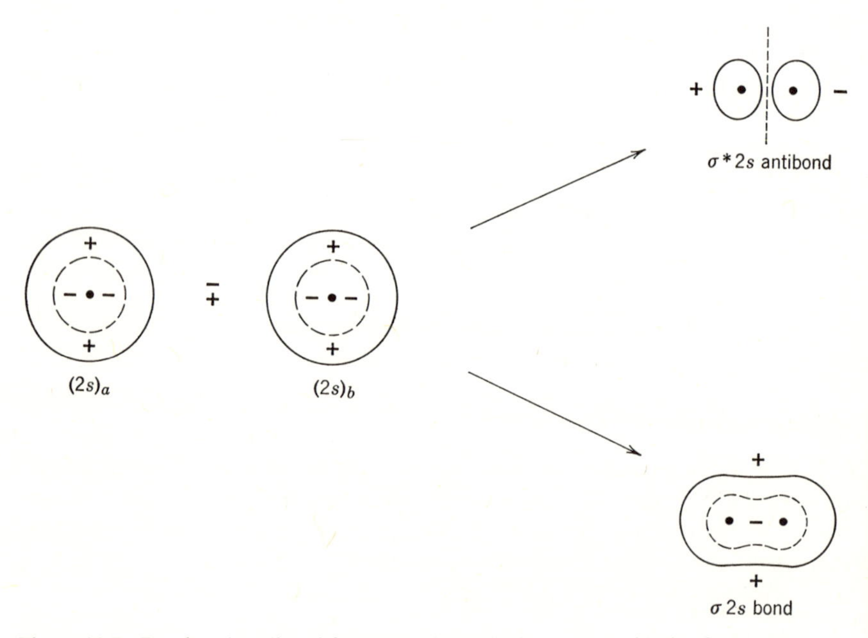

Figure 11.5 Bond and antibond formation from the $2s$ atomic orbitals of hydrogen.

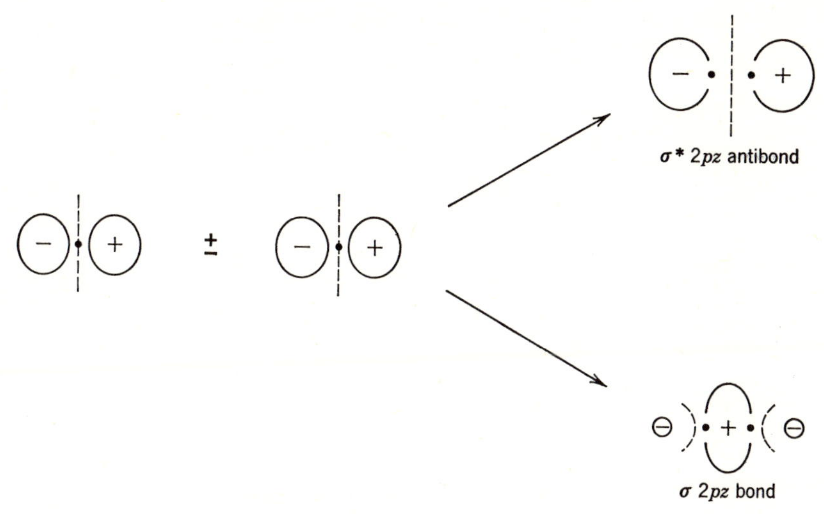

Figure 11.6 Bond and antibond formation from the $2pz$ atomic orbitals of hydrogen.

The designations "bonding" and "antibonding" may be a little puzzling here, for by synthesizing molecular orbitals out of wave functions for an excited hydrogen atom, the resulting energy for the $\sigma 2s$ bonding state must surely lie above the ground state energy of the infinitely separated nuclei? This is true, of course, and we may only call $\sigma 2s$ bonding by comparing its energy with that of a proton and an *excited* hydrogen atom at infinite separation.

Linear superposition of two $2pz$ atomic orbitals also leads to molecular orbitals of the σ type. The patterns are those of Figure 11.6, in which the bonding function is shown to have nodes which lie athwart the internuclear axis, but in positions beyond the nuclei, so that a buildup of electron density between the nuclei is still characteristic of a $\sigma 2pz$ molecular orbital. In passing the student should note that the bonding function is here synthesized from the *difference* of the two $2pz$ atomic orbitals rather than from their sum, since the symmetry of the atomic orbitals demands that their difference be nodeless in the region between the nuclei.

11.4 STATES OF π AND π* SYMMETRY

Further superposition of the patterns of atomic orbitals can be devised by using a couple of $2px$ functions. This is done in Figure 11.7, yielding molecular orbitals of π and π^* symmetry. The bonding $\pi 2px$ function exhibits a ribbon

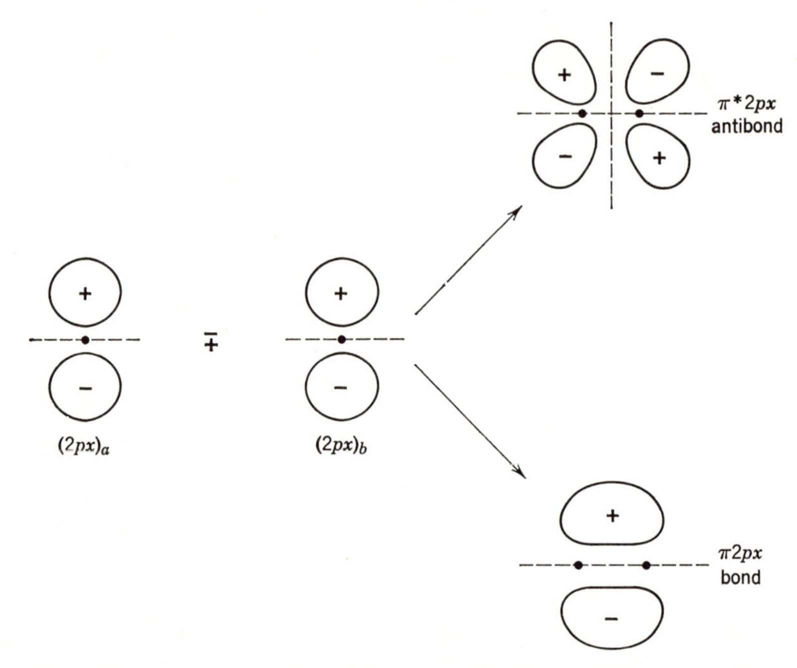

Figure 11.7 Bonding and antibonding molecular orbitals of π symmetry.

of charge above and below the nodal yz plane, but the bond energy calculated by comparing the energy of an electron in a $\pi 2px$ molecular orbital with that of an electron in the $2px$ atomic orbital of hydrogen is less than the bond energy of the $\pi 2pz$ state. We can understand this fact by comparing the overlap integrals

$$\int (2pz)_a(2pz)_b \, d\tau > \int (2px)_a(2px)_b \, d\tau$$

for a comparison of Figures 11.6 and 11.7 shows that the $2px$ functions do not interpenetrate nearly as efficiently as do the $2pz$ functions at a given internuclear separation, and we have learned to interpret a smaller overlap as implying a weaker bond. For the same reason, the π^*2px antibond is not as strongly antibonding as is the σ^*2pz antibond.

Because the $2px$ functions are identical with the $2py$ functions except for a rotation through 90°, we may expect equivalent $\pi 2py$ and π^*2py molecular orbitals exhibiting a node in the xz plane. Their energies will be identical with those of the $\pi 2px$ and π^*2px functions, and this confirms the results of Section 11.1 that π and π^* states are each doubly degenerate.

11.5 AN ENERGY LEVEL CHART FOR H_2^+

Upon evaluation of the necessary integrals, it turns out that the LCAO method predicts a sequence of energy states for H_2^+ which takes the form of Figure 11.8. I have shown, of course, only the LCAO wave functions derived from linear combinations of pairs of atomic orbitals up to the atomic $2p$ level. Each pair of atomic levels is correlated in the molecule with the bonding and antibonding levels into which it splits. The splitting is most

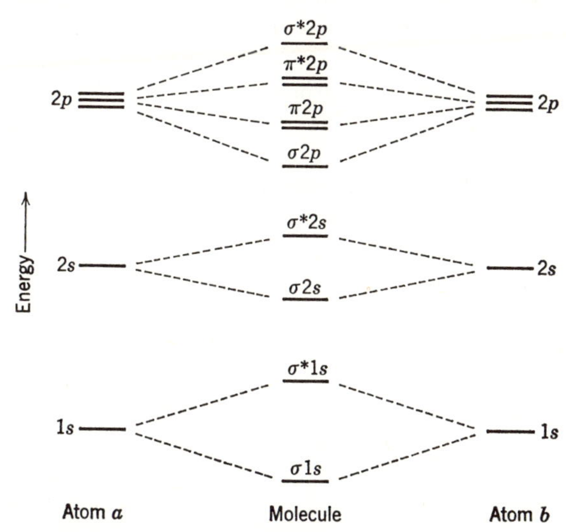

Figure 11.8 Energy level diagram for the H_2^+ ion.

complicated for the six atomic $2p$ functions (three on each center) which split into four levels, two of which retain double degeneracy. For hydrogen atoms the chart is incorrect in that it shows the $2s$ and $2p$ atomic states already separated in energy, but because the electrons of more complicated atoms can be considered to move in a non-Coulomb central field, the chart in its present form is better adapted to our subsequent generalizations.

Figure 11.8 is also incorrect in another way, for it shows the splitting of the atomic energy levels into bonding and antibonding states to be symmetric in the sense that the bonding level is lowered in energy by the same amount as the antibonding level is raised. This feature is a result of neglecting overlap S in equations 11.8 and 11.9 (see also equations 10.22 and Figure 10.9). If overlap is not neglected, the calculated molecular levels become unsymmetric, with the starred molecular orbitals stronger antibonds than the unstarred molecular orbitals are bonds.

11.6 HOMONUCLEAR DIATOMIC MOLECULES

In the world of homonuclear diatomic molecules, the H_2^+ ion plays a role similar to that enjoyed by the H atom in the periodic table. It is the one electron prototype structure whose wave functions and energy levels are an indispensable guide to the interpretation of complicated, polyelectronic structures. The reader will recall from Chapter 8 that, armed with a knowledge of the hydrogen atomic orbitals, our zeroth order approach to helium was to use the independent electron approximation and calculate approximate energy levels via a perturbation technique. Guided by this analogy, it would seem appropriate to write for the ground state of the hydrogen molecule H_2 an independent electron configuration $(\sigma 1s)^2$ in which two electrons are presumed to occupy the lowest bonding molecular orbital of H_2^+. If we take care to write a properly antisymmetrized wave function including spin, the ground state for H_2 will be predicted to be a singlet state with the spins paired, in agreement with experiment.

To continue our *aufbau* into higher members of a "periodic table" of homonuclear diatomics, we imagine a pair of neighboring helium nuclei upon which are cast three electrons to make He_2^+. All of the integrals α, β, S will change abruptly, for the greater nuclear charge of helium will cause its atomic wave functions to contract more closely about the nuclei than was the case for hydrogen. Despite this alteration in the calculated values of all the one electron energy levels, however, the relative sequence of levels needed to describe the ground state does not change, for which according to the Pauli principle we should have a configuration $(\sigma 1s)^2(\sigma^* 1s)$. With the inclusion of spin, the ground state should be doubly degenerate, for the $\sigma^* 1s$ electron could have either spin up or spin down. As for the stability of the structure, we need to balance the energy contribution of the two electrons in a bonding molecular orbital against the energy of a single electron in an antibonding one. The overall energy balance proves to be favorable, which means that He_2^+ is predicted to be stable, in agreement with experiment. To conform with conventional chemical practice, if a pair of electrons in a bonding molecular orbital represents a single chemical bond, then H_2^+ possesses a half bond, H_2 a single bond, and He_2^+ a half bond, with the $\sigma^* 1s$ electron cancelling out half of the $(\sigma 1s)^2$ single bond. Adding a fourth electron to make He_2 results in zero bonds, also in agreement with experiment.

These hints are very strong, and the student is invited to continue populating the bonding and antibonding levels of Figure 11.8 all the way through to Ne_2 (20 electrons), at which point all of the levels in Figure 11.8 are occupied with two electrons, and the overall number of chemical bonds is again zero. An *aufbau* based upon H_2^+ is, however, not entirely reliable, for experiment indicates some reordering of the levels as more and more electrons are added. Thus for N_2 (14 electrons), we should expect all levels through $\pi 2p$ to be completely filled, yielding a total of three bonds. Chemical experience supports the high stability of N_2 with a bond energy of 226 kcal/mole and a bond distance of 1.09 Å, but the evidence from ultraviolet spectroscopy is that the highest filled level in N_2 is $\sigma 2p$ rather than $\pi 2p$. This can only mean that the perturbing effect of interelectronic interaction between the fourteen electrons has been sufficient to invert the order of the $\sigma 2p$ and $\pi 2p$ levels as they appear in Figure 11.8. The student will recall a similar inversion in the atomic hydrogen sequence, where as our *aufbau* proceeds from argon to potassium, the extra electron goes into the 4s level of potassium rather than into the 3d level. At the present time the whole question of the exact order of the sequence . . . σ^*2s, $\sigma 2p$, $\pi 2p$, π^*2p, . . . is not entirely clear in the series of homonuclear diatomic molecules synthesized from the second row elements.[10,11]

Despite this problem, the addition of two electrons to the fourteen of N_2 moves us along to O_2 with a bond energy of 119 kcal/mole and a bond distance of 1.21 Å. The bond has evidently been weakened in comparison with N_2, and this is neatly explained from Figure 11.8 when the two new electrons are observed to be lodged in the antibonding π^*2p level. Oxygen is therefore predicted to have a double bond, but even more significant is the fact that when Hund's rule is applied to the two electrons in the doubly degenerate π^*2p level, we conclude that the effect of electronic interaction will be such as to force the molecule into a triplet state with each electron singly occupying one of the π^*2p levels and spins parallel. This simple explanation of the observed paramagnetism of oxygen was an early triumph of molecular orbital theory.

PROBLEMS

11.1 The bond energy and bond length of the fluorine molecule F_2 are 37 kcal/mole and 1.42 Å, respectively. Interpret these data by comparison with the same figures for O_2 and N_2.

11.2 Which of the species O_2, O_2^+, O_2^{2+} would be expected to be the most stable chemically? Which of the species N_2, N_2^+, N_2^{2+}? Justify your answers on the basis of molecular orbital theory.

11.3 The NO molecule is known to form a stable ion NO^+ in such compounds as nitrosyl perchlorate $NOClO_4$ and nitrosyl hydrogen sulfate $NOHSO_4$. On the basis of molecular orbital theory account for the relative ease with which NO forms a positive ion.

11.4 Predict whether beryllium vapor will be monatomic or diatomic.

11.5 The C_2 molecule is observed spectroscopically in flames where it is responsible for the blue color of the inner Bunsen cone. Assign electrons to the molecular orbitals of C_2 and predict whether the molecule will be paramagnetic or diamagnetic.

11.6 (1) Set up the Hamiltonian for the H_2 molecule using ∇^2 for the Laplacian operator, the letters a and b for the nuclei, and the numerals 1, 2 for the electrons. Let the internuclear distance be R.

(2) By dropping the electron repulsion term from the Hamiltonian, separate the simplified Schrödinger equation into two equations in the coordinates of a single electron only.

(3) Including spin, write an antisymmetric, zero order (independent electron), normalized wave function for the ground state of H_2 assuming that the exact wave function $\sigma 1s$ for the ground state of H_2^+ is known. How many antisymmetric wave functions are there for the ground state of H_2?

11.7 INNER SHELL AND VALENCE SHELL ELECTRONS

In the foregoing the student will have received the impression that by assigning all of the electrons in a homonuclear diatomic molecule to molecular orbitals, all of them participate equally in the chemical bonding. This is contrary to much chemical experience, which states that it is only those electrons in the outer or valence shell of an atom which are involved in molecule formation. The difficulty may be resolved by considering the effect of varying internuclear distance and nuclear charge upon the strength of the bonds formed by the various molecular orbitals. Thus in H_2 with a bond distance of 0.74 Å, the overlap integral between two $1s$ atomic orbitals centered on each nucleus is $S = 0.753$. This large overlap leads to considerable interaction between the orbitals; the resonance integral β which, according to Section 10.9, is the principal contributor to the bonding energy, is also large. For the lithium diatomic Li_2, the bond distance has increased to 2.67 Å and the increased charge on the nucleus has shrunken the effective radius of the $1s$ orbital so that the overlap integral of the $1s$ functions is reduced to $S = 0.000$. The contribution of the $1s$ electrons to the bonding in Li_2 is thus effectively zero and the charge distribution of two electrons in either the $\sigma 1s$ or $\sigma^* 1s$ molecular orbitals of Li_2 is hardly distinguishable from that of electrons in two totally separated atomic orbitals. Diagrams expressing the decrease in overlap and hence in interaction between $1s$ orbitals in H_2 and Li_2 are fancifully drawn in Figure 11.9. It follows that molecular orbital

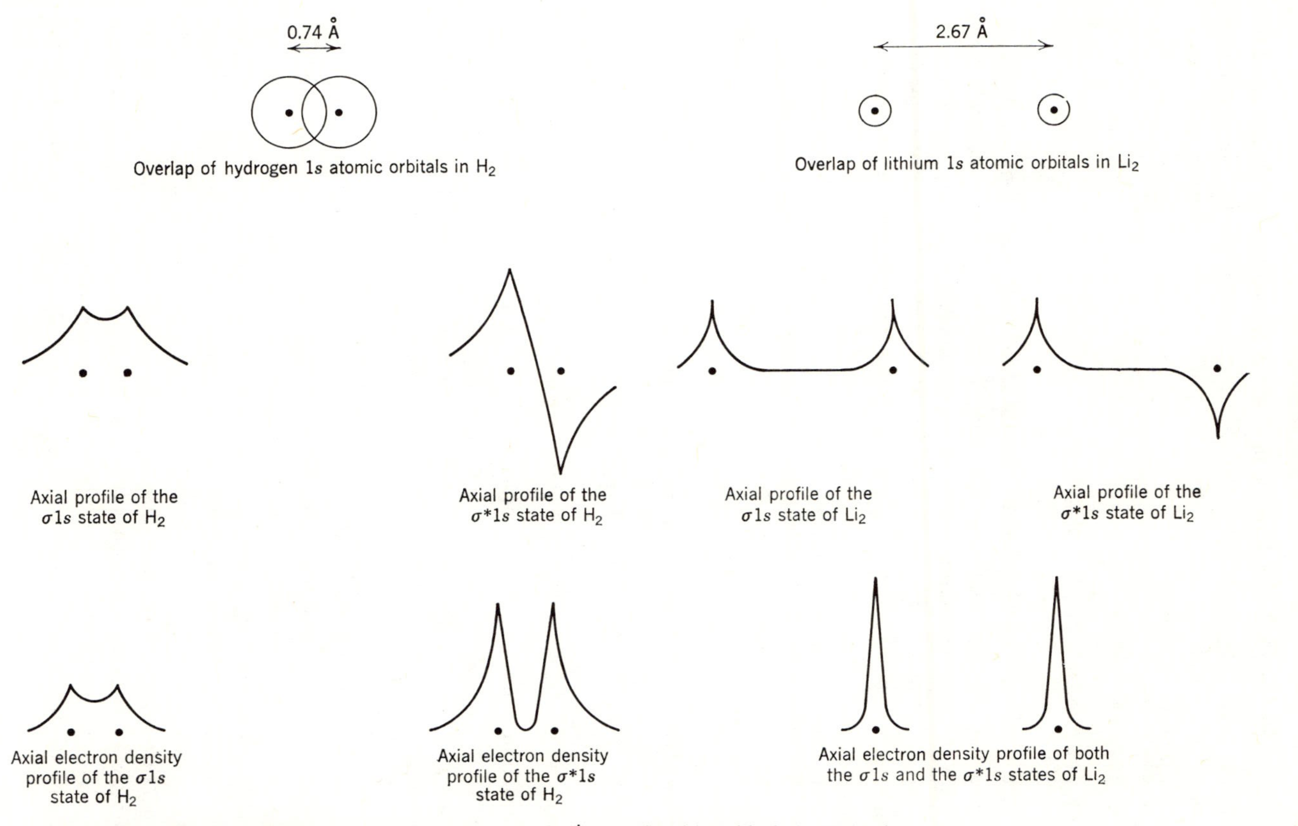

Figure 11.9 Effect of reduced overlap on the $\sigma 1s$ and $\sigma^* 1s$ molecular orbitals in H_2 and Li_2.

theory is not in any basic disagreement with the traditional conclusion of the chemist regarding the importance of valence electrons, for the inner shell molecular orbitals attenuate into atomic orbitals when there is only a small overlap between them.

11.8 THE NONCROSSING RULE

Let us reconsider what has by now become a tediously familiar problem: the roots of secular equations. Through all these many pages the student can hardly fail to have become aware of the extraordinary importance of secular equations in practical quantum mechanical calculations, and it is useful to pause at this point to emphasize some interesting properties illustrated by a simple 2×2 secular determinant.

Suppose in a molecular problem that we are led to a secular equation

$$\begin{vmatrix} \alpha_1 - E & \beta \\ \beta & \alpha_2 - E \end{vmatrix} = 0 \tag{11.10}$$

in which α_1 and α_2 are not necessarily equal. The roots are

$$\begin{aligned} E_1 &= \tfrac{1}{2}(\alpha_1 + \alpha_2) + \tfrac{1}{2}\sqrt{(\alpha_1 - \alpha_2)^2 + 4\beta^2} \\ E_2 &= \tfrac{1}{2}(\alpha_1 + \alpha_2) - \tfrac{1}{2}\sqrt{(\alpha_1 - \alpha_2)^2 + 4\beta^2} \end{aligned} \tag{11.11}$$

For a homonuclear diatomic molecule, equation 11.10 is the result of using a two term variation function and then neglecting overlap in the secular equation. For this special case, $\alpha_1 = \alpha_2 =$ approximately the energy of an electron in the atomic orbital of an isolated atom, and

$$E_1 = \alpha + \beta; \qquad E_2 = \alpha - \beta$$

so that β is a correction to the atomic energy states brought about by the presence of another atom in the neighborhood of the first.

When we come to apply the LCAO method to heteronuclear diatomic molecules, we may anticipate that α_1 and α_2, being the energies of an electron in the two different atomic orbitals centered on two different atoms, will no longer be equal (see Problem 2.6 and Section 6.5). Whatever the interpretation of α_1, α_2, however, it is always true that the roots of the molecular problem straddle the two roots of the atomic problem obtained for $\beta = 0$. The meaning of this statement is illustrated in Figure 11.10. At infinite separation of the nuclei, β is zero and the roots of the secular equation are from (11.11) the atomic energies α_1 and α_2. As the nuclei are brought together, β begins to depart from zero, and the new (bonding) root E_1 lies below either of the

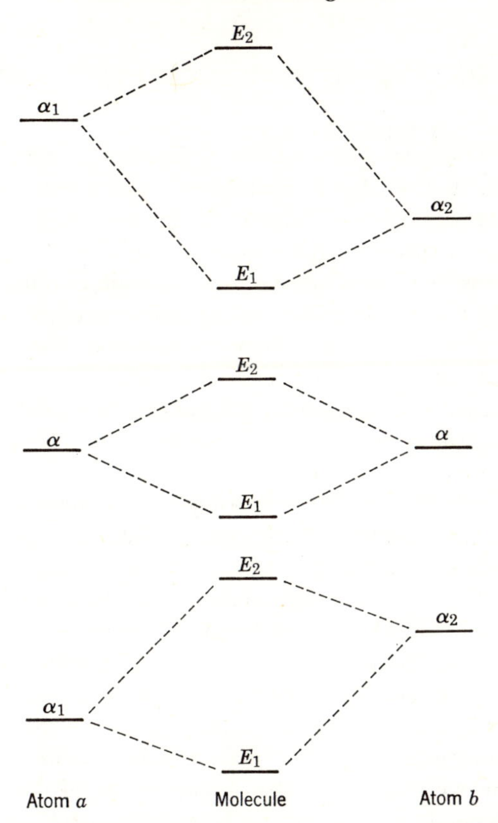

Figure 11.10 Modification of atomic energy levels as a result of molecule formation. In the center diagram the molecule is homonuclear. In the first diagram atom *b* is more electronegative than atom *a*. In the last diagram the reverse is the case.

atomic levels while the antibonding root E_2 lies above both of the atomic levels. It is as though the atomic energy levels were forced apart when the molecule is formed. The roots furthermore never cross in the sense that if we could continuously alter the charge on the nuclei so that atom *b* started out more electronegative than atom *a* ($\alpha_2 < \alpha_1$) and then we followed the sequence of changes in E_1 and E_2 as charge was continuously leaked from *b* to *a*, we should find that the difference $E_2 - E_1$ decreases to a minimum when the nuclei are identical and then increases again as their roles are reversed with *a* more electronegative than *b* ($\alpha_1 < \alpha_2$). The roots therefore never cross, and this is a first example of the noncrossing rule. As we come to improve our facility with the LCAO method, we shall relate this rule to considerations of the symmetry character of the molecular orbitals available to a molecule.

11.9 HETERONUCLEAR DIATOMIC MOLECULES AND THE IONIC BOND

The imaginary process of leaking charge from one nucleus to the other is a conceptually useful one, for it enables us to correlate the homonuclear pictures of Figures 11.5 to 11.7 with the corresponding patterns for heteronuclear diatomic molecules. As charge is transferred, the mathematical result in equations 11.11 is to make α_1 and α_2 begin to differ more and more from each other. If we were to follow the changing coefficients c_1 and c_2 in our LCAO variation wave function, we would find that the bonding state tended to weight the molecular orbital more and more in the direction of putting an increased electron density around the more electronegative atom, and that the antibonding state would do the opposite. The chemist would say in this case that the bond shows more and more ionic character, with electrons in the bonding orbital spending increasingly large amounts of time in the neighborhood of the more highly charged nucleus. The limiting situation is that the molecular orbital becomes simply an atomic orbital centered on the electronegative atom, and we have just seen from equations 11.11 and Figure 11.10 that in this extreme case the bonding energy level is identically the atomic energy α of the most electronegative atom. The electron has, in short, been completely transferred and the bond is completely ionic.

The weighting of the bonding molecular orbital toward the most electronegative of its constituent atomic orbitals and the simultaneous convergence of the molecular energy towards the atomic energy of this constituent is a matter which depends on the relative magnitudes of β^2 and $(\alpha_1 - \alpha_2)^2$ in equations 11.11. In the example chosen in the preceding paragraph, we imagined this process to be brought about by increasing the size of $|\alpha_1 - \alpha_2|$ relative to $|\beta|$, but the same effect will occur if β is made small. In our examples so far, β is small only if the internuclear distance is large, but it turns out that for certain combinations of atomic orbitals β is zero at *all* internuclear distances. While this result may appear surprising, we shall shortly have examples of it, and if it happens, we say that a pair of atomic orbitals ϕ_a, ϕ_b for which

$$\beta = \int \phi_a H \phi_b \, d\tau = 0$$

does not "combine" in molecular orbital formation.

11.10 HYBRIDIZATION

The student should remember at this point that the variation method is a way of guessing the solutions of Schrödinger's equation. So far we have used the LCAO approximation to guess at the contours of molecular orbitals, but we have consistently limited ourselves to the use of two atomic orbitals only, and this does not allow much room for maneuver in the search for greater accuracy. It is not unreasonable to expect that we shall obtain better results if we assume that each molecular orbital is a superposition of more than two of the constituent atomic orbitals, and this does, in fact, turn out to be the case.

Let us search for the best homonuclear, diatomic molecular orbitals which can be constructed as a linear combination of the *four* functions $(2s)_a$, $(2pz)_a$, $(2s)_b$, $(2pz)_b$. By choosing our basis set from atomic orbitals of principal quantum number 2, we have thrown emphasis on the formation of diatomic species from Li_2 through F_2; for in accordance with the discussion of Section 11.7, the valence electrons of an atom are those which make the dominant contribution to its chemical bonds. Writing

$$\psi = c_1(2s)_a + c_2(2pz)_a + c_3(2s)_b + c_4(2pz)_b$$

the variation formalism leads to a secular equation

$$\begin{vmatrix} \alpha_1 - E & \beta_3 & \beta_1 & \beta_4 \\ \beta_3 & \alpha_2 - E & \beta_4 & \beta_2 \\ \beta_1 & \beta_4 & \alpha_1 - E & \beta_3 \\ \beta_4 & \beta_2 & \beta_3 & \alpha_2 - E \end{vmatrix} = 0 \tag{11.12}$$

in which I have neglected all overlap integrals and

$$H_{11} = H_{33} = \int (2s)_a H (2s)_a \, d\tau \equiv \alpha_1$$

$$H_{22} = H_{44} = \int (2pz)_a H (2pz)_a \, d\tau \equiv \alpha_2$$

$$H_{12} = H_{34} = \int (2s)_a H (2pz)_a \, d\tau \equiv \beta_3$$

$$H_{13} = \int (2s)_a H (2s)_b \, d\tau \equiv \beta_1$$

$$H_{14} = H_{23} = \int (2s)_a H (2pz)_b \, d\tau \equiv \beta_4$$

$$H_{24} = \int (2pz)_a H (2pz)_b \, d\tau \equiv \beta_2$$

Of these integrals, α_1, α_2, β_1, β_2 are like integrals we have had before in Section 11.3; for they are the Coulomb and resonance integrals which arose when we used a two term variation function for our variation approximation. The integrals β_3 and β_4 are different, involving resonance interaction between

$2s$ and $2pz$ functions centered respectively on the same or on different nuclei.

What I am now about to do will seem to be unmotivated, but it is representative of operations with which we shall become increasingly familiar. The roots of secular determinants are not changed in value if rows are added or subtracted to other rows, or if columns are added or subtracted to other columns. The roots of determinant (11.12) will therefore not be affected if I replace it with another determinant whose first and third columns are respectively the sum and difference of columns 1 and 3 of (11.12); and whose second and fourth columns are the respective sum and difference of columns 2 and 4 of (11.12).

$$\begin{vmatrix} \alpha_1 + \beta_1 - E & \beta_3 + \beta_4 & \beta_1 - \alpha_1 + E & \beta_4 - \beta_3 \\ \beta_3 + \beta_4 & \alpha_2 + \beta_2 - E & \beta_4 - \beta_3 & \beta_2 - \alpha_2 + E \\ \alpha_1 + \beta_1 - E & \beta_3 + \beta_4 & \alpha_1 - \beta_1 - E & \beta_3 - \beta_4 \\ \beta_3 + \beta_4 & \alpha_2 + \beta_2 - E & \beta_3 - \beta_4 & \alpha_2 - \beta_2 - E \end{vmatrix} = 0 \quad (11.13)$$

Now treat the rows of (11.13) in a similar way, adding and subtracting rows 1 and 3 and rows 2 and 4. After factoring out and discarding a common factor 2 the result is

$$\begin{vmatrix} \alpha_1 + \beta_1 - E & \beta_3 + \beta_4 & 0 & 0 \\ \beta_3 + \beta_4 & \alpha_2 + \beta_2 - E & 0 & 0 \\ 0 & 0 & \alpha_1 - \beta_1 - E & \beta_3 - \beta_4 \\ 0 & 0 & \beta_3 - \beta_4 & \alpha_2 - \beta_2 - E \end{vmatrix} \quad (11.14)$$

$$= \begin{vmatrix} \alpha_1 + \beta_1 - E & \beta_3 + \beta_4 \\ \beta_3 + \beta_4 & \alpha_2 + \beta_2 - E \end{vmatrix} \times \begin{vmatrix} \alpha_1 - \beta_1 - E & \beta_3 - \beta_4 \\ \beta_3 - \beta_4 & \alpha_2 - \beta_2 - E \end{vmatrix} = 0$$

Our juggling of rows and columns has thus led to a factorization of the original characteristic polynomial of order 4 into a pair of quadratics.

Suppose now that in equation 11.14 we set $\beta_3 = \beta_4 = 0$. Physically this amounts to neglecting all resonance interaction between $2s$ and $2pz$ atomic orbitals. Mathematically the energy levels simplify to $E = \alpha_1 \pm \beta_1$; $\alpha_2 \pm \beta_2$ which are precisely of the form (11.9) after S has been neglected. Such interactions were implicitly ignored when in Sections 11.3 to 11.5 we used only a two term basis in our variation calculation. With β_3 and β_4 not equal to zero, however, the noncrossing rule of secular equations comes into play, and the resulting approximate energy levels will be such as to lower the energy of the bonding states and to raise the energy of the antibonding states. Recall the variation theorem that the true ground state energy of a quantum mechanical system must be less than or equal to any energy calculated by an approximate method, and we see that the inclusion of interaction between $2s$ and $2pz$ atomic orbitals results in an improvement in the resulting molecular orbitals.

When the coefficients c in the LCAO wave functions are calculated, the unnormalized molecular orbital of lowest energy has the appearance

$$\sigma_1 = (2s)_a + \gamma(2pz)_a - \gamma(2pz)_b + (2s)_b \tag{11.15}$$

in which γ is a coefficient which varies with internuclear distance, becoming rapidly zero as $R \to \infty$. If we make the approximation $\beta_3 = \beta_4 = 0$, then $\gamma = 0$ at all distances, and we recover exactly the form (11.9) of the $\sigma 2s$ bonding wave function. Equation 11.15 with $\gamma \neq 0$ thus represents an improvement of the lowest σ orbital obtainable as an LCAO of atomic orbitals of principal quantum number 2.

A linear combination of two or more atomic orbitals centered on the same atom is known as an *atomic hybrid*. The pattern of the hybrid $(2s) + \gamma(2pz)$

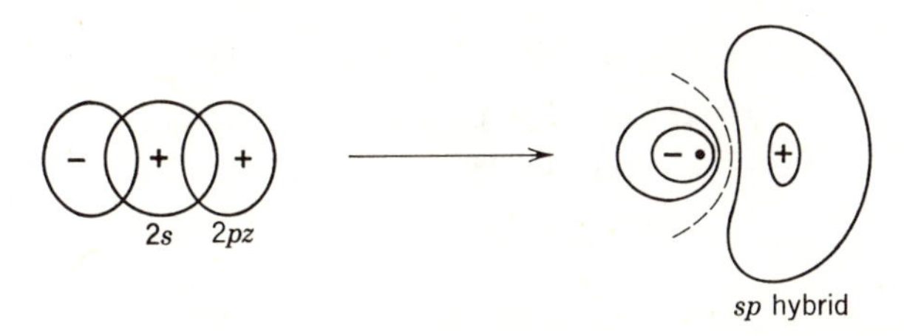

Figure 11.11 Construction of an unnormalized *sp* hybrid $(2s) + (2pz)$.

can be most easily pictured for the special case $\gamma = 1$ (Figure 11.11). We observe that the *sp* hybrid atomic orbital has better directional properties than either the *s* or *p* atomic orbitals alone. If the sign of γ is reversed, $\gamma = -1$, the hybrid $(2s) - (2pz)$ points in the opposite direction, and equation 11.15 is nothing other than a superposition of a hybrid $(2s) + \gamma(2pz)$ centered on a plus a hybrid $(2s) - \gamma(2pz)$ centered on b. Because the overlapping hybrids increase the electron density between the nuclei beyond that described by the simpler superposition of atomic $2s$ orbitals alone, the calculated bond energy is improved.

Hybridization thus arises naturally whenever we seek an improvement in our molecular orbitals by letting more and more atomic orbitals interact. It becomes significant, however, only if at least two of the Coulomb integrals α which appear on the diagonal of the secular equation are roughly equal. That is, if the energies of an electron in either of two atomic orbitals located *on the same atom* are α_1 and α_2, then for $|\alpha_1 - \alpha_2|$ large, there is very little hybridization between these atomic orbitals. This property of secular equations assures us that hybridization does not occur between $1s$ orbitals and $2p$ orbitals located on the same atom. It will, in fact, occur significantly only for the closely spaced energy levels occupied in whole or in part by the valence electrons of the atom.

11.11 OTHER SIGMA HYBRIDS

We have just seen that a variation function using two atomic orbitals on each atom leads to a better approximation of the bonding sigma molecular orbital of lowest energy in a homonuclear, diatomic molecule. The other roots of the 4×4 secular equation 11.14 are approximations to the energy levels of the other sigma molecular orbitals, both bonding and antibonding. When the complete set of coefficients c of the LCAO molecular orbitals is calculated we obtain four functions which, unnormalized and in order of increasing energy, are

$$\begin{aligned} \sigma_1 &= (2s)_a + \gamma(2pz)_a - \gamma(2pz)_b + (2s)_b \\ \sigma_2 &= (2s)_a + \gamma(2pz)_a + \gamma(2pz)_b - (2s)_b \\ \sigma_3 &= -\gamma(2s)_a + (2pz)_a - (2pz)_b - \gamma(2s)_b \\ \sigma_4 &= -\gamma(2s)_a + (2pz)_a + (2pz)_b + \gamma(2s)_b \end{aligned} \tag{11.16}$$

If in the secular equation 11.14 we put $\beta_3 = \beta_4 = 0$, we should automatically have $\gamma = 0$ also, and the linear combinations (11.16) reduce to those of Section 11.3 which we labeled $\sigma 2s$, $\sigma^* 2s$, $\sigma 2p$, $\sigma^* 2p$. By arbitrarily setting $\gamma = 1$ the patterns of Figure 11.12 are obtained. These patterns are recognizably the offspring of those of Figures 11.5 and 11.6.

The unnormalized combination $(2s) + \gamma(2p)$ occurs so frequently in LCAO molecular orbitals set up for second row elements that special names have been given to hybrids for special values of γ. If $\gamma = \pm 1$, we have already identified the linear combination as an *sp* hybrid. If $\gamma = \pm\sqrt{2}$, it is known as an sp^2 hybrid, and if $\gamma = \pm\sqrt{3}$, it is an sp^3 hybrid. The reader has been familiar with the geometry of these hybrids since his freshman course. All of them have enhanced directional character over either purely s or purely p atomic orbitals.

It is often stated in chemical texts that the creation of hybrid orbitals in, say, carbon, requires the expenditure of energy because the valence shell configuration $(2s)^2(2p)^2$ of carbon must first be replaced by the excited state $(2s)(2p)^3$ before hybridization can take place, but that the energy absorbed is more than compensated by the extra stability of the bonds formed. Without wishing to destroy the picturesqueness of this language, it is

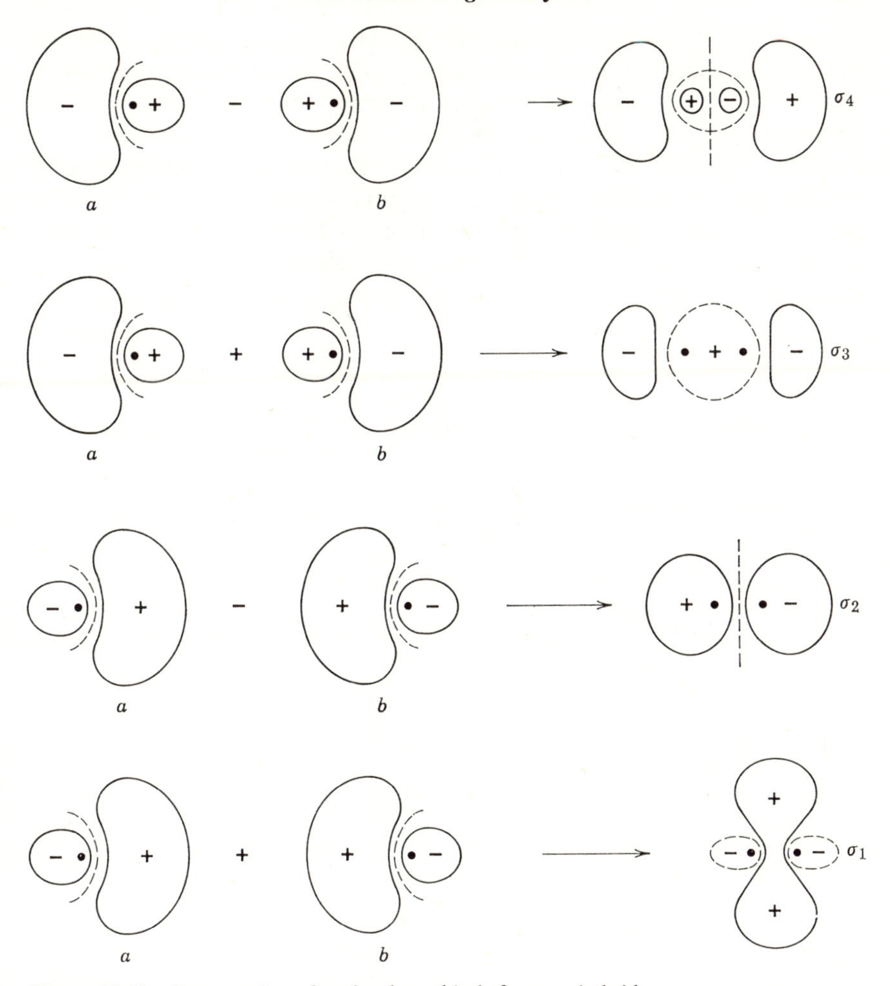

Figure 11.12 Construction of molecular orbitals from *sp* hybrids.

instructive to identify its mathematical origin. Let

$$\alpha_1 = \int (2s)H(2s)\,d\tau$$

$$\alpha_2 = \int (2p)H(2p)\,d\tau$$

be the Coulomb integrals respectively for the $2s$ and $2p$ atomic orbitals of carbon and remember that we may interpret α_1 and α_2 as approximately the energies of electrons resident in these atomic states. The difference $\alpha_2 - \alpha_1$ is thus an approximation to the promotional energy needed to excite an electron from $2s$ to $2p$. Now if $|\alpha_2 - \alpha_1|$ is small compared to the resonance integrals $|\beta|$ which occur in the secular equation of our LCAO calculation, then hybridization will be encouraged. Conversely the larger $|\alpha_2 - \alpha_1|$ with respect to the $|\beta|$'s the less of this type of orbital mixing will occur. The situation is mathematically identical with the difference between purely covalent and purely ionic bonds; for if α_1 and α_2 are Coulomb integrals for atomic orbitals centered on *different* atoms, then the discussion of Section 11.9 shows that $|\alpha_2 - \alpha_1|$ large implies small mixing of the orbitals and a largely ionic bond.

REFERENCES

1. H. Eyring, J. Walter, and G. E. Kimball, *Quantum Chemistry*, John Wiley and Sons, Inc., New York, 1944, p. 367.
2. W. Kauzmann, *Quantum Chemistry*, Academic Press, New York, 1957, p. 379.
3. H. Margenau and G. M. Murphy, *The Mathematics of Physics and Chemistry*, Vol. 1, D. Van Nostrand, Inc., New York, 1956, pp. 180–182.
4. H. Eyring, J. Walter, and G. E. Kimball, *loc. cit.*, pp. 201–203.
5. W. Kauzmann, *loc. cit.*, pp. 226–227.
6. L. Pauling and E. B. Wilson, Jr., *Introduction to Quantum Mechanics*, McGraw-Hill, Inc., New York, 1935, pp. 333–340.
7. F. L. Pilar, *Elementary Quantum Chemistry*, McGraw-Hill, Inc., New York, 1968, pp. 456–467.
8. H. Eyring, J. Walter, and G. E. Kimball, *loc. cit.*, pp. 196–197.
9. W. Kauzmann, *loc. cit.*, 379–380.
10. G. Herzberg, *Molecular Spectra and Molecular Structure*, D. Van Nostrand, Inc., New York, 1950, pp. 323–333.
11. G. W. King, *Spectroscopy and Molecular Structure*, Holt, Rinehart and Winston, Inc., New York, 1964, pp. 130–131.

Chapter 12

SYMMETRY: LOCALIZED AND NONLOCALIZED MOLECULAR ORBITALS

12.1 SIGMA AND PI SYMMETRY IN DIATOMICS

Certain very important results with a far reaching effect on the nature of bond formation in molecules come out of a study of the symmetry of atomic orbitals as they appear in LCAO molecular orbitals. We have seen that by forming (unnormalized) LCAO's of two atomic orbitals apiece

$$(2s)_a \pm (2s)_b; \qquad (2pz)_a \pm (2pz)_b$$

we obtained our first approximations to the sigma bonds and antibonds of a homonuclear diatomic molecule. Instead of taking two linear combinations of two functions at a time we obtained better results in Sections 11.10 and 11.11 if we used four linear combinations of all four atomic orbitals together, and the fruit of this union was improved accuracy plus an awareness of the importance of hybridization. Why could not this program be continued by mixing in $2px$ and $2py$ atomic orbitals along with the $2s$ and $2pz$ we have already mixed so as to improve the accuracy still further? Remember that the $2px$ and $2py$ functions led in Section 11.4 to our first picture of the π bond, so that the present program would be to consider both σ and π type molecular orbitals together.

An important theorem concerning symmetry prevents progress from being achieved in this direction, and we shall show that atomic orbitals that have different symmetry classifications will not mix together in an LCAO molecular orbital. In particular, σ and π type molecular orbitals stay rigidly separate from each other.

In any diatomic molecule (not necessarily homonuclear) consider the separate orbitals centered on the different atoms before we blend them into an LCAO. If, for example, our variation function contains $(2s)_a$, $(2pz)_a$, $(2s)_b$, and $(2pz)_b$, then we have seen that the resonance integrals

$$\beta_3 = \int (2s)_a H (2pz)_a \, d\tau$$

$$\beta_4 = \int (2s)_a H (2pz)_b \, d\tau$$

occur in our secular equation, as also to be accurate do the overlap integrals

$$S_3 = \int (2s)_a (2pz)_a \, d\tau$$

$$S_4 = \int (2s)_a (2pz)_b \, d\tau$$

It is the magnitudes of these β's and S's that determine the relative contribution of the $2s$ and the $2pz$ functions to the final LCAO, and if we arbitrarily set $\beta_3 = \beta_4 = S_3 = S_4 = 0$, then the $2s$ will not "mix" or hybridize with the $2pz$ functions, that is, the final LCAO's will contain either $2s$ or $2pz$ but not both.

Suppose now that we incorporate both $(2px)_a$ and $(2px)_b$ into our variation wave function. Then the secular equation will contain additional integrals of the type

$$\begin{aligned} \beta_5 &= \int (2s)_a H (2px)_a \, d\tau; \qquad S_5 = \int (2s)_a (2px)_a \, d\tau \\ \beta_6 &= \int (2s)_a H (2px)_b \, d\tau; \qquad S_6 = \int (2s)_a (2px)_b \, d\tau \end{aligned} \tag{12.1}$$

in off diagonal positions in the secular determinant. *The quantities* (12.1) *are exactly zero*. This is not an approximation and is not dependent upon the internuclear distance R. It is mathematically exact.

To see how this comes about consider the overlap integral S_6. The integrand is the product of the two atomic functions sketched in Figure 12.1, one of which has the property of being invariant upon reflection through the yz plane, while the other does not change in magnitude but does change in sign. The total effect of the reflection operation is thus to change the sign of the integrand. But the integration process simply sums up volume elements $d\tau$ weighted by the factor $(2s)_a(2px)_b$; and for every $d\tau$ which enters with a positive weight, there is another $d\tau$ on the other side of the yz plane which enters with an equal but negative weight. The cumulative result of the integration is thus zero, nor does this fact change with internuclear distance so long as the reflection symmetry through yz is preserved.

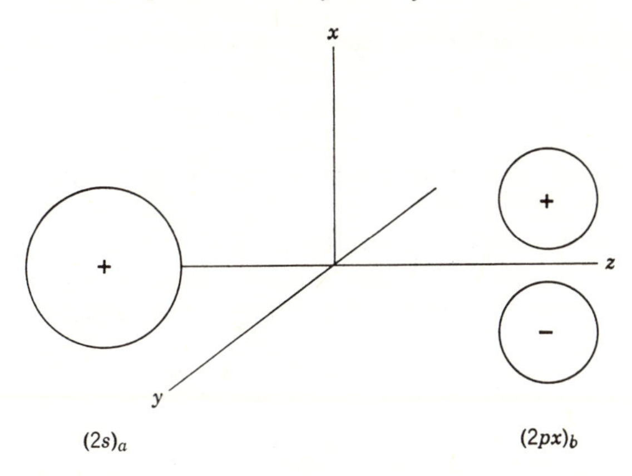

Figure 12.1 Atomic orbitals of different symmetry type.

A similar argument holds for the integral β_6. The Hamiltonian operator H and particularly its potential energy part V has the symmetry of the molecule. We had examples of this in connection with "bond" formation for the isotropic oscillator in Section 6.11 and again in our discussion of equations 11.3 for the H_2^+ ion. A reflection through yz leaves H invariant because changing the sign of x does not alter the potential energy V. Of the three factors in the integrand of β_6, reflection therefore leaves $(2s)_a$ and H unchanged, but it changes the sign of $(2px)_b$. The entire integrand is consequently antisymmetric about the yz plane and $\beta_6 = 0$.

A generalization of this argument to any type of symmetry in a molecule leads to the conclusion that an off diagonal element (resonance or overlap integral) of a secular equation always vanishes whenever the atomic orbitals of its integrand belong to different symmetry classifications. "Symmetry classification" here means a set of atomic orbitals, or linear combinations of atomic orbitals, all of which are transformed into linear combinations of members of the same set when subjected to one of the symmetry operations of the molecule, that is, some reflection or rotation of the molecule which brings it into an orientation physically indistinguishable from its initial orientation.

In the example chosen above, the symmetry operation was reflection through the yz plane of a diatomic molecule. The $2s$ and $2pz$ atomic orbitals have σ symmetry and are unaffected by the reflection. As a result the integrals

$$\beta_3 = \int (2s)_a H (2pz)_a \, d\tau; \qquad \beta_4 = \int (2s)_a H (2pz)_b \, d\tau$$

do not vanish, and hybrids form between atomic orbitals $2s$ and $2pz$ centered on the same atom. On the other hand, the $2px$ function has π or antisymmetry when subjected to this reflection, and in consequence the cross product

$$\beta_6 = \int (2s)_a H(2px)_b \, d\tau = 0$$

meaning that $2s$ and $2px$ orbitals will not mix in an LCAO for a diatomic molecule.

Going back to the original plan of blending all s and p atomic orbitals for a diatomic molecule into one giant variation function, we perceive that a secular equation constructed along these lines will have the structure

$$\left|\begin{array}{c|c} \sigma \text{ class} & \begin{matrix} 0 & \cdots & 0 \\ \vdots & & \vdots \\ 0 & \cdots & 0 \end{matrix} \\ \hline \begin{matrix} 0 & \cdots & 0 \\ \vdots & & \vdots \\ 0 & \cdots & 0 \end{matrix} & \pi \text{ class} \end{array}\right| = 0$$

in which integrals comprising the first block are of the type

$$\beta = \int (\sigma \text{ atomic orbital}) H (\sigma \text{ atomic orbital}) \, d\tau$$

$$S = \int (\sigma \text{ atomic orbital})(\sigma \text{ atomic orbital}) \, d\tau$$

those of the second block are of the type

$$\beta = \int (\pi \text{ atomic orbital}) H (\pi \text{ atomic orbital}) \, d\tau$$

$$S = \int (\pi \text{ atomic orbital})(\pi \text{ atomic orbital}) \, d\tau$$

and the zeros are the cross products between symmetry classes.

$$\int (\sigma \text{ atomic orbital}) H (\pi \text{ atomic orbital}) \, d\tau = 0$$

$$\int (\sigma \text{ atomic orbital})(\pi \text{ atomic orbital}) \, d\tau = 0$$

The secular equation therefore factors into two polynomials, the roots of the first yielding energy levels of exclusively σ states and the roots of the second yielding only π states. The noncrossing rule will then operate to keep separate from each other the energy levels of σ states and to keep separate from each other the energy levels of π states, but the predicted relative positions in the overall energy scale of a σ molecular orbital and a π molecular orbital can and do change as a result of improvements in the variation function or of changing internuclear distance.

We have already observed this phenomenon in atoms, for which, of course, symmetry arguments are equally valid. An atom possesses perfect reflection symmetry through any plane including the nucleus and perfect rotational symmetry around any axis through the nucleus. The sequence of symmetry classes is $s, p, d, \ldots$; for every s function is invariant under every symmetry operation of the atom, every p function is sent into a linear combination of other p functions by the same operations, every d into a linear combination of d's, and so on. The noncrossing rule then acts to assure us that the sequence of s energy levels will always be $1s, 2s, 3s, \ldots$ with no inversions in the order; that the sequence of p energy levels will always be $2p, 3p, \ldots$; and so on. There is, however, nothing in the noncrossing rule that prevents an inversion in the order of, say, a d state with an s state; and this does, in fact, happen. In the helium atom, for instance, $3d$ lies lower in energy than $4s$, but in the iron atom the order of these levels is reversed.

12.2 LOCALIZED MOLECULAR ORBITALS; THE WATER MOLECULE

We have so far considered only diatomic molecules and have observed that chemical bond formation takes place whenever the valence electrons of atoms can find energy states lower in the presence of two nuclei together rather than two nuclei separate. When we come to extend these ideas to polyatomic molecules, we find that it is useful to distinguish between localized and nonlocalized chemical bonds. Examples of the former type are the familiar C—C and C—H single bonds which thermochemical and infared data suggest are very little different in character as they appear in thousands of different compounds. An example of the latter type are the giant molecular bonds formed in aromatic compounds in which we have good reason to believe that the π bonds extend their charge distribution over many different atoms.

To illustrate localized bonds, let us study the water molecule. The geometry of H_2O is familiar from Figure 12.2. I place the oxygen atom at the

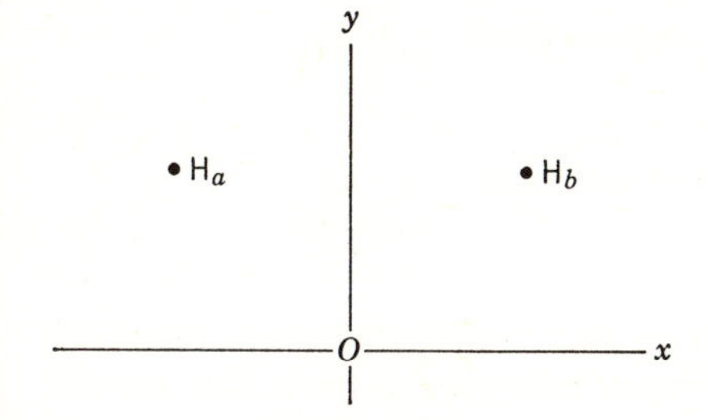

Figure 12.2 Coordinate system for the water molecule.

origin of an xy coordinate system with the hydrogen atoms making 45° angles with the x and y axes. This is, of course, not the correct bond angle observed in the molecule, but to simplify the algebra while emphasizing the physical ideas, I shall use only the unhybridized $2px$ and $2py$ valence orbitals of oxygen, for which maximum overlap with the hydrogen $1s$ orbitals occurs at a bond angle of 90°. This means that two of the valence electrons of oxygen are assigned to the $2s$ level and two more are assigned to the $2pz$ level, leaving two left over for bonding purposes.

The naïvest approach, and one that from the outset makes the chemical bond localized, is to direct $2p$ orbitals from the oxygen atom toward each

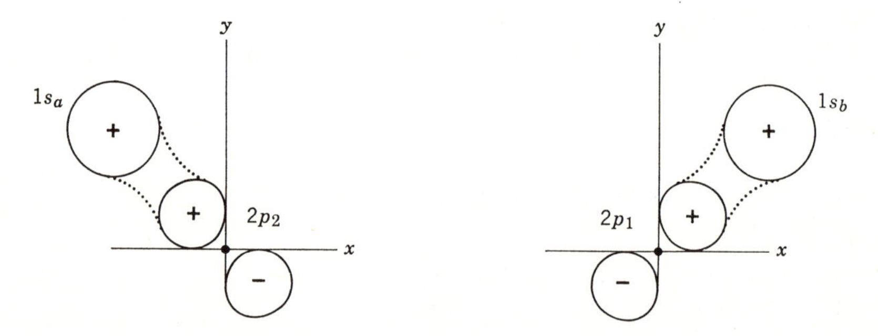

Figure 12.3 Localized bond formation in water.

hydrogen atom and to let them couple in *diatomic* molecular orbitals with the $1s$ hydrogen atomic orbitals. Each of the O—H bonds thus formed is constructed by ignoring the presence of the other hydrogen atom (Figure 12.3), and into each of the resulting bonding σ molecular orbitals we feed two electrons.

In algebraic detail, the $2p$ oxygen orbitals needed are constructed by rotating the degenerate pair $2px$ and $2py$ through 45°. This is accomplished by inverting the familiar orthogonal rotation matrix of Figure 1.6:

$$(2p)_1 = \frac{1}{\sqrt{2}}[(2px) + (2py)]$$

$$(2p)_2 = \frac{1}{\sqrt{2}}[-(2px) + (2py)]$$

The student would do well to convince himself either algebraically or by the vectorial superposition of the px and py directional characters that $(2p)_1$ and $(2p)_2$ have the forms sketched in Figure 12.3. Ignoring normalization and the different electronegativities of hydrogen and oxygen, the bonding molecular orbitals will be qualitatively

$$\sigma_1 = (1s)_b + (2p)_1 = (1s)_b + \frac{1}{\sqrt{2}}[(2px) + (2py)]$$
$$\sigma_2 = (1s)_a + (2p)_2 = (1s)_a + \frac{1}{\sqrt{2}}[-(2px) + (2py)] \quad (12.2)$$

Because of their symmetry we anticipate that σ_1 and σ_2 will be degenerate.

Now let us try to set up some nonlocalized molecular orbitals whose synthesis does not possess the aesthetically unattractive feature of ignoring the hydrogen atoms one at a time. Instead we shall consider both hydrogen atoms simultaneously on an equal footing. This may be accomplished by taking as a variation wave function a linear combination of all four atomic orbitals simultaneously rather than the two linear combinations (12.2) of three at a time. We could start with

$$\psi = c_1(1s)_a + c_2(1s)_b + c_3(2px) + c_4(2py) \quad (12.3)$$

and then find the c's by the variation method, but it is easier to proceed by trickery, making use of the symmetry of the molecule.

Note that the water molecule is symmetric about the yz plane and that the $2py$ orbital is invariant upon reflection through this plane. The $2px$ orbital, however, changes sign under the reflection operation, whence we conclude that $2px$ and $2py$ belong to different symmetry classes.

The hydrogen $1s$ orbitals by themselves are neither symmetric nor antisymmetric upon reflection, being sent into each other. In a manner analogous to the construction of the functions (8.4) in our study of the Pauli principle, however, of the orthogonal linear combinations

$$\frac{1}{\sqrt{2}}[(1s)_a + (1s)_b]$$
$$\frac{1}{\sqrt{2}}[(1s)_a - (1s)_b]$$

the first is symmetric and the second antisymmetric upon reflection. We possess therefore a symmetric class of basis functions

$$(2py), \frac{1}{\sqrt{2}}[(1s)_a + (1s)_b]$$

and an antisymmetric class

$$(2px), \frac{1}{\sqrt{2}}[(1s)_a - (1s)_b]$$

Should we write our variation function in terms of this basis

$$\psi = c_1(2py) + c_2 \frac{1}{\sqrt{2}}[(1s)_a + (1s)_b] + c_3(2px) + c_4 \frac{1}{\sqrt{2}}[(1s)_a - (1s)_b] \tag{12.4}$$

we would find that (12.4) has the advantage over (12.3) of a factored secular equation,

$$\begin{vmatrix} \alpha_1 - E & \sqrt{2}\,\beta_1 & 0 & 0 \\ \sqrt{2}\,\beta_1 & \alpha_2 + \beta_3 - E & 0 & 0 \\ 0 & 0 & \alpha_3 - E & \sqrt{2}\,\beta_2 \\ 0 & 0 & \sqrt{2}\,\beta_2 & \alpha_2 - \beta_3 - E \end{vmatrix} = 0 \tag{12.5}$$

In (12.5) all overlap integrals have been neglected and

$$\alpha_1 = \int (2py)H(2py)\,d\tau$$

$$\alpha_2 = \int (1s)_a H(1s)_a\,d\tau = \int (1s)_b H(1s)_b\,d\tau$$

$$\alpha_3 = \int (2px)H(2px)\,d\tau$$

$$\beta_1 = \int (2py)H(1s)_a\,d\tau = \int (2py)H(1s)_b\,d\tau$$

$$\beta_2 = \int (2px)H(1s)_a\,d\tau = -\int (2px)H(1s)_b\,d\tau$$

$$\beta_3 = \int (1s)_a H(1s)_b\,d\tau$$

The construction of basis sets of orbitals that are organized into symmetry classes always results in a factored secular equation, one factor for each symmetry class. The factorization could have been accomplished directly from the secular equation derived from (12.3) by juggling rows and columns after the manner of equations 11.12 to 11.14, but the construction of symmetry bases is a much surer and quicker way.

Let us now make some further approximations. The hydrogen atoms in water are much further apart than they are in either H_2 or H_2^+. It is therefore reasonable to expect that to a good approximation $\beta_3 = 0$. We have furthermore learned to interpret a Coulomb integral as approximately an atomic energy state, whence α_1 and α_3 must be very nearly the same. For the same reason the similar orientation of $2px$ and $2py$ with respect to $(1s)_a$ must make the equality $\beta_1 = \beta_2$ roughly valid. With these approximations equation 12.5 becomes

$$\begin{vmatrix} \alpha_1 - E & \sqrt{2}\,\beta_1 & 0 & 0 \\ \sqrt{2}\,\beta_1 & \alpha_2 - E & 0 & 0 \\ 0 & 0 & \alpha_1 - E & \sqrt{2}\,\beta_1 \\ 0 & 0 & \sqrt{2}\,\beta_1 & \alpha_2 - E \end{vmatrix} = 0 \tag{12.6}$$

whence by inspection the roots of the secular equation are seen to be doubly degenerate. Without actually performing the necessary numerical work, we sense qualitatively that the bonding pair of degenerate orbitals will be those linear combinations within each symmetry class which place no nodes between the oxygen and hydrogen atoms

$$\begin{aligned} \sigma_3 &= \frac{1}{\sqrt{2}}[(1s)_a + (1s)_b] + (2py) \\ \sigma_4 &= -\frac{1}{\sqrt{2}}[(1s)_a - (1s)_b] + (2px) \end{aligned} \tag{12.7}$$

The structure of σ_3 and σ_4 is shown geometrically in Figure 12.4. These degenerate molecular orbitals are nonlocalized, that is, an electron placed in one of them runs over the whole molecule. They thus differ sharply in appearance from the localized molecular orbitals σ_1 and σ_2 of Figure 12.3 which require an electron to spend all of its time along the bond axis between the oxygen and one of the hydrogen atoms.

At first sight this would appear to be a distressing result, and one has the impression that one can derive anything one pleases from quantum mechanics. The localized and nonlocalized pictures for water are, however, quite reconcilable, and understanding emerges when we use our molecular orbitals

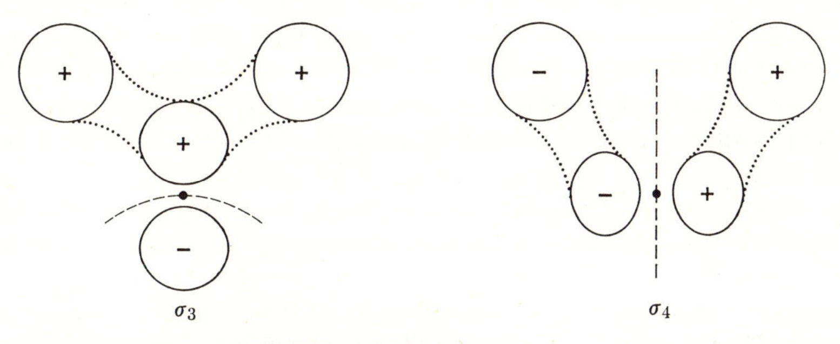

Figure 12.4 Nonlocalized bond formation in water.

to construct antisymmetrized, independent electron wave functions for the four valence electrons involved in the bonding:

$$\psi(1, 2, 3, 4) = \frac{1}{\sqrt{4!}} \begin{vmatrix} \sigma_3(1)\alpha(1) & \sigma_3(2)\alpha(2) & \sigma_3(3)\alpha(3) & \sigma_3(4)\alpha(4) \\ \sigma_3(1)\beta(1) & \sigma_3(2)\beta(2) & \sigma_3(3)\beta(3) & \sigma_3(4)\beta(4) \\ \sigma_4(1)\alpha(1) & \sigma_4(2)\alpha(2) & \sigma_4(3)\alpha(3) & \sigma_4(4)\alpha(4) \\ \sigma_4(1)\beta(1) & \sigma_4(2)\beta(2) & \sigma_4(3)\beta(3) & \sigma_4(4)\beta(4) \end{vmatrix}$$

This function corresponds to the configuration $(\sigma_3)^2(\sigma_4)^2$ for four bonding electrons in the two nonlocalized molecular orbitals. In an analogous way we write for the configuration $(\sigma_1)^2(\sigma_2)^2$ in the localized picture

$$\psi(1, 2, 3, 4) = \frac{1}{\sqrt{4!}} \begin{vmatrix} \sigma_1(1)\alpha(1) & \sigma_1(2)\alpha(2) & \sigma_1(3)\alpha(3) & \sigma_1(4)\alpha(4) \\ \sigma_1(1)\beta(1) & \sigma_1(2)\beta(2) & \sigma_1(3)\beta(3) & \sigma_1(4)\beta(4) \\ \sigma_2(1)\alpha(1) & \sigma_2(2)\alpha(2) & \sigma_2(3)\alpha(3) & \sigma_2(4)\alpha(4) \\ \sigma_2(1)\beta(1) & \sigma_2(2)\beta(2) & \sigma_2(3)\beta(3) & \sigma_2(4)\beta(4) \end{vmatrix}$$

Now despite the different appearance of these four-electron wave functions, they are identical. To establish this fact, the student should first satisfy himself from equations 12.2 and 12.7 that the apparently different pairs (σ_1, σ_2), (σ_3, σ_4) are simply orthogonal linear combinations of each other:

$$\begin{aligned} \sigma_1 &= \frac{1}{\sqrt{2}}(\sigma_3 + \sigma_4) \\ \sigma_2 &= \frac{1}{\sqrt{2}}(\sigma_3 - \sigma_4) \end{aligned} \tag{12.8}$$

He may also want to check this fact geometrically by superposing the patterns of Figure 12.4 according to the recipe (12.8) so as to recover the patterns of Figure 12.3. It now follows that our two determinants above are identical because the rows of the second are orthogonal linear combinations of the rows of the first, and it is known that such juggling of rows does not change the value of a determinant. Explicitly, the first row of the second determinant is the sum of rows 1 plus 3 of the first (divided by $\sqrt{2}$), and with this hint the student is invited to complete the transformation of the first determinant into the second.

To summarize, in the water molecule both an approximate localized description of the bonds and an approximate nonlocalized one lead to the same results. Whichever description is used, both predict that between each hydrogen and the oxygen is the equivalent in electron density of one pair of electrons. The Pauli principle prevents us from knowing which pair of electrons is in a particular bond. But while the motion of an individual electron cannot be followed, the total charge density of the four bonding electrons is always such as to make the bonds equivalent to two electrons apiece. The traditional and instinctive response of the chemist to this fact is to call the *bonds* localized even if the *electrons* are not.

Finally, the identity of the localized and nonlocalized descriptions of water is the result of the approximations which led from equation 12.5 to equation 12.6. If our level of accuracy is such that $\alpha_1 \neq \alpha_3$ or $\beta_3 \neq 0$, and so on, then we are including in our calculation small effects due, for example, to the mutual interaction of the hydrogen atoms. It is these small effects that account for the fact that the O—H bond energy or infared absorption frequency is not perfectly constant in series of different compounds containing the OH group. The bonds are thus very nearly localized, but not quite.

PROBLEMS

12.1 If the $2s$ orbital of oxygen were included in the basis set for a nonlocalized LCAO calculation for water, to which symmetry class would it belong? Sketch the form of the 5×5 secular equation showing its expected factorization.

12.2 Set up symmetry classes for water for the six basis functions $2s$, $2px$, $2py$, $2pz$, $(1s)_a$, $(1s)_b$. In addition to reflection symmetry through the yz plane, include the symmetry operation of reflection through the xy plane.

12.3 An sp^n hybrid atomic orbital is a linear combination of s and p hydrogen-like atomic orbitals (e.g., Problem 5.6)

$$(sp^n) = a(s) + b(px) + c(py) + d(pz)$$

such that $a^2 + b^2 + c^2 + d^2 = 1$ and $n = (b^2 + c^2 + d^2)/a^2$. Note that n need not be an integer.

The directional character of the hybrid is determined solely by the constituent p orbitals, the hybrid pointing in the direction $\mathbf{v} = (b, c, d)$ in three dimensional space. Thus two orthogonal sp^2 hybrids lying in the xy plane, pointing into the upper half plane, and symmetrically disposed with respect to the y axis are

$$(sp^2)_1 = \frac{1}{\sqrt{3}}(s) + \frac{1}{\sqrt{2}}(px) + \frac{1}{\sqrt{6}}(py) \qquad \text{with} \quad \mathbf{v}_1 = \left(\frac{1}{\sqrt{2}}, \frac{1}{\sqrt{6}}, 0\right)$$

$$(sp^2)_2 = \frac{1}{\sqrt{3}}(s) - \frac{1}{\sqrt{2}}(px) + \frac{1}{\sqrt{6}}(py) \qquad \text{with} \quad \mathbf{v}_2 = \left(-\frac{1}{\sqrt{2}}, \frac{1}{\sqrt{6}}, 0\right)$$

The bond angle θ between $(sp^2)_1$ and $(sp^2)_2$ is determined from the coefficients of the p orbitals:

$$\cos \theta = \frac{\mathbf{v}_1 \cdot \mathbf{v}_2}{|v_1|\,|v_2|} = -\tfrac{1}{2}$$

so that $\theta = 120°$.

In water the H—O—H bond angle is known experimentally to be 104.5°. In the xy plane construct two orthonormal sp^n hybrids that are identical except for orientation, point into the upper half plane, are symmetrically disposed with respect to the y axis, and exhibit a bond angle of 104.5°. Find also the value of n for these hybrids.

12.4 The hybrids of Problem 12.3 reproduce the correct bond angle for water. Construct a third hybrid in the xy plane

$$(sp^n) = a(s) + b(px) + c(py) + d(pz)$$

which is normalized and orthogonal to *both* of the hybrids of Problem 12.3. In what direction does this third hybrid point? Would you expect electrons occupying this hybrid to be bonding or nonbonding? Would they contribute to the dipole moment of the water molecule? Note that n for the third hybrid is not the same as n for the hybrids of Problem 12.3.

12.3 HÜCKEL MOLECULAR ORBITAL THEORY REVISITED: BUTADIENE

The sprinkling of α's and β's throughout this and the preceding chapter cannot but have convinced the reader that the Hückel molecular orbital theory of Chapter 2 has something to do with the LCAO variation method. Let us derive the necessary formalism for butadiene and identify the approximations which led to the secular determinant (2.1).

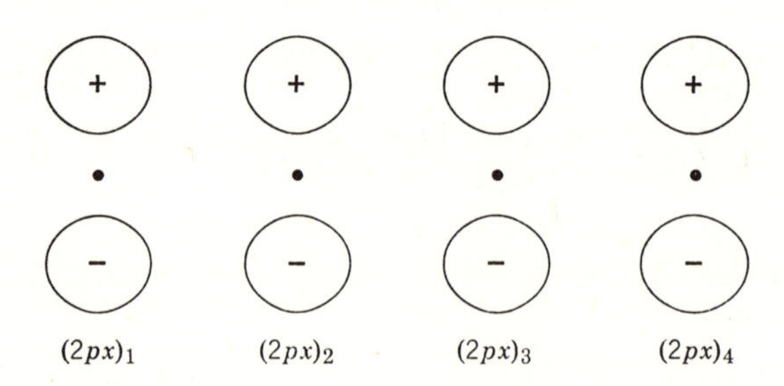

Figure 12.5 Atomic orbitals for the π bonds in butadiene.

We restrict ourselves to a consideration of the π network only and picture the $2px$ atomic orbitals of the four carbon atoms as perpendicular to the yz plane (Figure 12.5). The different reflection symmetry with respect to the molecular plane of the π network compared with the σ network permits us to write LCAO's involving the $2px$ orbitals only:

$$\psi = c_1(2px)_1 + c_2(2px)_2 + c_3(2px)_3 + c_4(2px)_4$$

A straightforward application of the variation method leads to a secular equation

$$|H_{ij} - ES_{ij}| = 0$$

in which

$$H_{ij} = \int (2px)_i H(2px)_j \, d\tau$$

$$S_{ij} = \int (2px)_i (2px)_j \, d\tau$$

It is a good approximation to take all the Coulomb integrals to be the same for each carbon,

$$\alpha = \int (2px)_i H(2px)_i \, d\tau$$

and also to write

$$\beta = \int (2px)_1 H(2px)_2 \, d\tau$$

for the resonance integral between all nearest neighbors. Resonance integrals between nonnearest neighbors are set equal to zero because of the large internuclear distance involved. Similarly we write

$$S = \int (2px)_1 (2px)_2 \, d\tau$$

for overlap integrals between nearest neighbors and set all nonnearest neighbor overlap integrals equal to zero. The result is

$$\begin{vmatrix} \alpha - E & \beta - ES & 0 & 0 \\ \beta - ES & \alpha - E & \beta - ES & 0 \\ 0 & \beta - ES & \alpha - E & \beta - ES \\ 0 & 0 & \beta - ES & \alpha - E \end{vmatrix} = 0 \qquad (12.9)$$

Equation 12.9 becomes identical with (2.1) if we make the additional, bad approximation $S = 0$. The results of this approach, as we have seen, yield the energy levels as functions of α and β plus the coefficients $c_1, \ldots, c_4$ for the first four LCAO molecule orbitals. The bad approximation $S = 0$ is compensated to some extent by taking α and β to be empirical parameters, adjustable to the type of data of interest, say heat of hydrogenation data or alternatively ultraviolet absorption data.

Unlike water, the π network of butadiene cannot be described by localized bonds.* Interpreting literally the conventional formula $CH_2{=}CH{-}CH{=}CH_2$, we consider each olefinic bond separately and write a variation function

$$\psi_1 = c_1(2px)_1 + c_2(2px)_2$$

for the first and another

$$\psi_2 = c_3(2px)_3 + c_4(2px)_4$$

for the second without any interaction between the two. Each function leads to a secular determinant

$$\begin{vmatrix} \alpha - E & \beta \\ \beta & \alpha - E \end{vmatrix} = 0$$

and this is precisely what we should obtain from (2.1) if we were to strike out the 23 and 32 components and replace them with zeros. The approximation is arbitrary and a very poor one, and the difference between the two approaches led in Section 2.5 to the definition of a delocalization energy to account for the extra stability of delocalized π bonds over localized, olefinic ones. In sum, while localized and nonlocalized molecular orbitals can to a good approximation be reconciled for the water molecule, the two approaches for butadiene are irreconcilable, with the nonlocalized picture the more accurate. The overall electron density of two electrons per bond for water suggests the description "saturated" for these bonds, but butadiene with four π electrons in all does not distribute these four into a pair between each of the two remotest pairs of carbon atoms. The π network is to this extent "unsaturated" with partial double bond character at each bond position. It is the wish to express this partial double bond character quantitatively which led in Section 2.5 to the definition of bond order.

PROBLEMS

12.5 The butadiene molecule possesses rotational symmetry through an angle of 180° about an axis perpendicular to the yz plane at the midpoint of the 23 bond. Set up four orthogonal, linear combinations of

* This statement is true only in the sense that molecular properties cannot be represented as a sum of contributions from the set of *aliphatic* single bonds and *olefinic* double bonds drawn in the classical structure. M. J. S. Dewar[10] has shown, however, that conjugated polyenes have properties that are additive functions of the numbers of single and double bonds in the classical formula, provided that the additive parameters used are neither aliphatic nor olefinic.

the original basis set $(2px)_1, \ldots, (2px)_4$ which are either symmetric or antisymmetric under this rotation and show that the 4×4 secular equation is thereby factored into two 2×2 determinants. Derive the energy levels of the factored determinant and show that they are identically equations 2.2.

12.6 Interpreting the components of the Hückel vectors of equations 2.3 as the coefficients of $2px$ atomic orbitals centered on the several carbon atoms, sketch approximate contours for the four molecular orbitals showing nodes and the algebraic sign of the wave function.

12.7 A pair of bonding localized molecular orbitals for butadiene is

$$\pi_1 = \frac{1}{\sqrt{2}}[(2px)_1 + (2px)_2]$$

$$\pi_2 = \frac{1}{\sqrt{2}}[(2px)_3 + (2px)_4]$$

From equations 2.3 the pair of nonlocalized molecular orbitals of lowest energy is

$$\pi_3 = (0.372)(2px)_1 + (0.601)(2px)_2 + (0.601)(2px)_3 + (0.372)(2px)_4$$
$$\pi_4 = (0.601)(2px)_1 + (0.372)(2px)_2 - (0.372)(2px)_3 - (0.601)(2px)_4$$

Set up a four electron wave function in the form of a determinant for the localized configuration $(\pi_1)^2(\pi_2)^2$ and for the nonlocalized configuration $(\pi_3)^2(\pi_4)^2$. Can one determinant be transformed into the other by forming orthogonal linear combinations of the rows? What does this say about the validity of a localized description of the π bonds in butadiene?

12.4 A THREEFOLD AXIS OF SYMMETRY: TRIMETHYLENEMETHANE

To make maximum use of the symmetry properties of molecules in quantum mechanical investigations the powerful methods of group theory are required. I shall make no attempt to develop the abstract algebra of groups in any systematic way. Instead I shall tease the reader with an intuitive, "how to do it" example showing a slightly more complicated symmetry than any discussed up to this point.

Our work will be based on the Hückel diagram of Figure 12.6, a schematic representation of the trimethylenemethane diradical $CH_2{=}CH(CH_2\cdot)_2$. As usual, the π network is synthesized approximately out of LCAO's based on four atomic carbon $2px$ functions oriented perpendicular to the plane of the

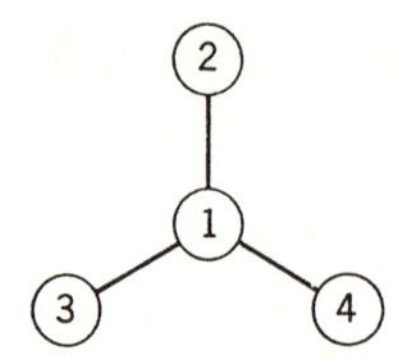

Figure 12.6 Hückel diagram for the trimethylene methane diradical.

molecule, $(2px)_1$, $(2px)_2$, $(2px)_3$, $(2px)_4$. If these atomic functions are used directly in a variation form

$$\psi = c_1(2px)_1 + c_2(2px)_2 + c_3(2px)_3 + c_4(2px)_4 \tag{12.10}$$

then the usual, unfactored, secular equation is obtained:

$$\begin{vmatrix} \alpha - E & \beta & \beta & \beta \\ \beta & \alpha - E & 0 & 0 \\ \beta & 0 & \alpha - E & 0 \\ \beta & 0 & 0 & \alpha - E \end{vmatrix} = 0 \tag{12.11}$$

The unfactored form of (12.11) is the result of using a basis set of functions that do not transform in a simple way under the symmetry operations of the molecule. Thus we observe that a rotation of 120° about an x axis passing through atom 1 brings the Hückel diagram into an orientation indistinguishable from the original diagram. The official group theoretical symbol for a clockwise rotation of 120° is C_3, in which the 3 means that the complete rotation of 360° is divided into three equal parts. Let us examine the effect of the operator C_3 upon each of our original basis functions:

$$\begin{aligned} C_3(2px)_1 &= (2px)_1 \\ C_3(2px)_2 &= (2px)_4 \\ C_3(2px)_3 &= (2px)_2 \\ C_3(2px)_4 &= (2px)_3 \end{aligned}$$

The central atomic orbital $(2px)_1$ transforms in a particularly simple way, for the rotation sends it into itself. The atomic orbitals centered on the peripheral atoms, however, are rotated into other members of the peripheral set. If we are ingenious, we observe that the linear combination

$$(2px)_2 + (2px)_3 + (2px)_4 \tag{12.12}$$

transforms in a manner identical with $(2px)_1$, for the rotation simply rearranges the terms in the sum. The normalized Hückel vectors for $(2px)_1$

and the symmetric linear combination (12.12) are, respectively

$$\psi_1 = (1, 0, 0, 0)$$

$$\psi_2 = \frac{1}{\sqrt{3}}(0, 1, 1, 1)$$

They are orthogonal and satisfy the operator equations

$$C_3\psi_1 = \psi_1$$

$$C_3\psi_2 = \psi_2$$

ψ_1 and ψ_2 are two orthonormal vectors in the vector space of four dimensions. If we are to have a complete basis for our variation calculation, we shall require two more orthonormal vectors, and the easiest way to

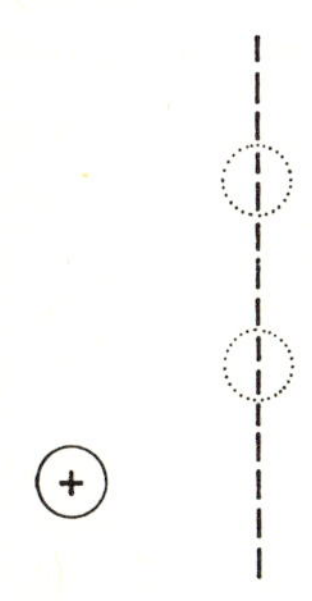

Figure 12.7 A trimethylene methane molecular orbital showing a node.

construct a third vector orthogonal to the first two is to run a node through the molecule (Figure 12.7). The Hückel vector displaying this node is

$$\psi_3 = \frac{1}{\sqrt{2}}(0, 0, 1, -1)$$

Let us see what happens to ψ_3 when operated on by C_3:

$$C_3\psi_3 = \frac{1}{\sqrt{2}}(0, 1, -1, 0)$$

meaning that the node has now been shifted to pass through atom 4. $C_3\psi_3$ is evidently not identical with ψ_3, so we let C_3 operate a second time, $C_3{}^2\psi_3 = (1/\sqrt{2})(0, -1, 0, 1)$. While $C_3{}^2\psi_3$ is not identical with either ψ_3 or $C_3\psi_3$, it can be expressed as a linear combination of them,

$$C_3{}^2\psi_3 = -\psi_3 - C_3\psi_3$$

and we have an example of a set of two vectors ψ_3 and $C_3\psi_3$ which successive applications of C_3 send into linear combinations of themselves.

While ψ_3 and $C_3\psi_3$ transform together as a set, they have the disadvantage of not being orthogonal to each other, and this motivates the construction of

$$\psi_4 = p\psi_3 + qC_3\psi_3$$

normalized and orthogonal to ψ_3. I leave it to the reader to discover that this requires $p = 1/\sqrt{3}$ and $q = 2/\sqrt{3}$, so that

$$\psi_4 = \frac{1}{\sqrt{6}}(0, 2, -1, -1)$$

and now we have two orthonormal vectors ψ_3 and ψ_4 which under rotation transform into linear combinations of themselves.

The possibilities for classification of molecular orbitals on the basis of the rotation operation are now exhausted, but trimethylene methane possesses other symmetry operations. A reflection through a plane including the 1–2 bond axis and perpendicular to the plane of the molecule leaves $(2px)_1$ and $(2px)_2$ unmodified but interchanges $(2px)_3$ and $(2px)_4$. The official group theoretical symbol for this reflection operation is σ_v, and we let σ_v operate on each of ψ_1, ψ_2, ψ_3, ψ_4 in turn:

$$\sigma_v\psi_1 = \psi_1$$
$$\sigma_v\psi_2 = \psi_2$$
$$\sigma_v\psi_3 = -\psi_3$$
$$\sigma_v\psi_4 = \psi_4$$

To be sure, there are two other reflection planes of symmetry in the molecule, but what we have already done is sufficient to accomplish the decomposition of (12.11) into factored form. Our set of four vectors falls into three symmetry classes. The first is the totally symmetric class $\{\psi_1, \psi_2\}$, which has the property that ψ_1 and ψ_2 are transformed into themselves by every symmetry operation of the molecule. The second class is ψ_3 all by itself, for while ψ_3 and ψ_4 transform together under rotation, they behave differently under reflection σ_v, and the members of a class must transform in the same way for *every* symmetry operation. The final class is ψ_4 all by itself.

Reverting to function space notation, the components of the Hückel vectors ψ are the coefficients of our basis functions φ to be used in a variation calculation,

$$\varphi_1 = (2px)_1$$

$$\varphi_2 = \frac{1}{\sqrt{3}}[(2px)_2 + (2px)_3 + (2px)_4]$$

$$\varphi_3 = \frac{1}{\sqrt{2}}[(2px)_3 - (2px)_4]$$

$$\varphi_4 = \frac{1}{\sqrt{6}}[2(2px)_2 - (2px)_3 - (2px)_4]$$

and from this basis we construct the variation form,

$$\psi = c_1\psi_1 + c_2\psi_2 + c_3\psi_3 + c_4\psi_4 \tag{12.13}$$

When we do this we shall find that resonance integrals taken between symmetry classes all vanish. Thus

$$H_{11} = \int \psi_1 H \psi_1 \, d\tau = \int (2px)_1 H (2px)_1 \, d\tau = \alpha$$

$$\begin{aligned} H_{22} &= \int \psi_2 H \psi_2 \, d\tau \\ &= \tfrac{1}{3}\int [(2px)_2 + (2px)_3 + (2px)_4] H [(2px)_2 + (2px)_3 + (2px)_4] \, d\tau \\ &= \tfrac{1}{3}(\alpha + \alpha + \alpha) = \alpha \end{aligned}$$

$$\begin{aligned} H_{12} &= \int \psi_1 H \psi_2 \, d\tau \\ &= \frac{1}{\sqrt{3}} \int (2px)_1 H [(2px)_2 + (2px)_3 + (2px)_4] \, d\tau \\ &= \frac{1}{\sqrt{3}}(\beta + \beta + \beta) = \sqrt{3}\,\beta \end{aligned}$$

but

$$\begin{aligned} H_{13} &= \int \psi_1 H \psi_3 \, d\tau \\ &= \frac{1}{\sqrt{2}} \int (2px)_1 H [(2px)_3 - (2px)_4] \, d\tau \\ &= \frac{1}{\sqrt{2}}(\beta - \beta) = 0 \end{aligned}$$

The student should work out the remaining integrals to find a factored secular equation

$$\begin{vmatrix} \alpha - E & \sqrt{3}\,\beta & 0 & 0 \\ \sqrt{3}\,\beta & \alpha - E & 0 & 0 \\ 0 & 0 & \alpha - E & 0 \\ 0 & 0 & 0 & \alpha - E \end{vmatrix} = 0 \tag{12.14}$$

with roots $E = \alpha + \sqrt{3}\beta, \alpha, \alpha, \alpha - \sqrt{3}\beta$.

PROBLEMS

12.8 Calculate the roots of equation 12.11 and compare them with those of (12.14).

12.9 From (12.14) calculate the coefficients c_j for the variation form (12.13) and convert to Hückel molecular orbitals of the form (12.10). Sketch the physical appearance of the molecular orbitals over the Hückel diagram of Figure 12.6, showing all nodes.

12.5 APOLOGY

This book ends at a point where from the chemist's point of view it is only beginning to beome interesting. My purpose, however, was not to take the student the whole way, but to prepare him for the different directions he may wish to go. One possible direction is backwards: a return to the foundation of our subject by reading an advanced text.[1–4] Or he may wish to progress to more detailed applications of quantum mechanics to problems in chemistry by reading more on molecular orbital theory[5–10] or more about symmetry.[11,12] To the student who will do neither and in whom the routine of higher education has not crushed the capacity to wonder, I leave an aesthetically haunting thought. The laws of Nature as revealed in quantum mechanics are geometrical laws, not in the space of three dimensions about which we learned in high school, but in an extension of the early perceptions of the Greeks to spaces of infinitely many dimensions. To the extent that they have been found useful in quantum mechanics, the necessary advances in geometry were complete by the close of the nineteenth century, in good time to be coordinated with the concurrent advances in experimental physics by a few brilliant minds. The reader has been a witness to the child of this union. My book is a success to the extent that he has been exhilarated by it.

REFERENCES

1. H. Eyring, J. Walter, and G. E. Kimball, *Quantum Chemistry*, John Wiley and Sons, Inc., New York, 1944.
2. W. Kauzmann, *Quantum Chemistry*, Academic Press, Inc., New York, 1957.
3. L. Pauling and E. B. Wilson, *Introduction to Quantum Mechanics*, McGraw-Hill Book Co., New York, 1935.
4. F. L. Pilar, *Elementary Quantum Chemistry*, McGraw-Hill, Inc., 1968.
5. A. Streitwieser, Jr., *Molecular Orbital Theory for Organic Chemists*, John Wiley and Sons, Inc., New York, 1961.

6. J. D. Roberts, *Notes on Molecular Orbital Calculations*, W. A. Benjamin, Inc., New York, 1962.
7. C. J. Ballhausen and H. B. Gray, *Molecular Orbital Theory*, W. A. Benjamin, Inc., New York, 1965.
8. C. A. Coulson, *Valence*, Oxford Press, London, 1961.
9. M. Orchin and H. H. Jaffé, *The Importance of Antibonding Orbitals*, Houghton Mifflin Co., Boston, 1967.
10. M. J. S. Dewar, *The Molecular Orbital Theory of Organic Chemistry*, McGraw-Hill Book Co., New York, 1969.
11. F. A. Cotton, *Chemical Applications of Group Theory*, Interscience, New York, 1963.
12. H. H. Jaffé and M. Orchin, *Symmetry in Chemistry*, John Wiley and Sons, New York, 1965.

Appendix I

PHYSICAL CONSTANTS TO FOUR SIGNIFICANT FIGURES

Electronic charge	e	4.803×10^{-10} esu ($g^{1/2}$ $cm^{3/2}$ sec^{-1})
Mass of the electron	m_e	9.109×10^{-28} g
Planck's constant	h	6.626×10^{-27} erg sec
$h/2\pi$	$\hbar$	1.055×10^{-27} erg sec
Boltzmann's constant	k	1.381×10^{-16} erg deg^{-1}
Avogadro's number	N	6.023×10^{23} $mole^{-1}$
Bohr radius $\hbar^2/m_e e^2$	a_0	0.5292 Å
Twice the ionization energy of the hydrogen atom $m_e e^4/\hbar^2$. This is also identical with the Coulomb electrostatic energy e^2/a_0 of two electronic charges separated by one Bohr radius.	E_0	627.5 kcal $mole^{-1}$
Velocity of light	c	2.998×10^{10} cm sec^{-1}

Appendix II

COORDINATE SYSTEMS

PLANE POLAR COORDINATES

$$x = r\cos\varphi$$

$$y = r\sin\varphi$$

$$\nabla^2 = \frac{1}{r}\frac{\partial}{\partial r}r\frac{\partial}{\partial r} + \frac{1}{r^2}\frac{\partial^2}{\partial\varphi^2}$$

$$d\tau = r\,dr\,d\varphi$$

$$0 \le r \le \infty; \qquad 0 \le \varphi \le 2\pi$$

SPHERICAL POLAR COORDINATES (see Figure 5.1)

$$x = r\sin\theta\cos\varphi$$

$$y = r\sin\theta\sin\varphi$$

$$z = r\cos\theta$$

$$\nabla^2 = \frac{1}{r^2}\frac{\partial}{\partial r}r^2\frac{\partial}{\partial r} + \frac{1}{r^2\sin\theta}\frac{\partial}{\partial\theta}\sin\theta\frac{\partial}{\partial\theta} + \frac{1}{r^2\sin^2\theta}\frac{\partial^2}{\partial\varphi^2}$$

$$d\tau = r^2\sin\theta\,d\theta\,d\varphi$$

$$0 \le r \le \infty; \qquad 0 \le \theta \le \pi; \qquad 0 \le \varphi \le 2\pi$$

Appendix III

NORMALIZED, HYDROGEN-LIKE WAVE FUNCTIONS

Let the nucleus be of atomic number Z and define $\rho = Zr/a_0$; $a = a_0/Z$ so that a is the Bohr radius for a nuclear charge Z and ρ is the radial coordinate measured in units of a.

$$(1s) = a^{-3/2} \frac{1}{\sqrt{\pi}} e^{-\rho}$$

$$(2s) = a^{-3/2} \frac{1}{4\sqrt{2\pi}} (2 - \rho)e^{-\rho/2}$$

$$(2px) = a^{-3/2} \frac{1}{4\sqrt{2\pi}} \rho e^{-\rho/2} \sin\theta \cos\varphi$$

$$(2py) = a^{-3/2} \frac{1}{4\sqrt{2\pi}} \rho e^{-\rho/2} \sin\theta \sin\varphi$$

$$(2pz) = a^{-3/2} \frac{1}{4\sqrt{2\pi}} \rho e^{-\rho/2} \cos\theta$$

$$(3s) = a^{-3/2} \frac{1}{81\sqrt{3\pi}} (27 - 18\rho + 2\rho^2)e^{-\rho/3}$$

$$(3px) = a^{-3/2} \frac{\sqrt{2}}{81\sqrt{\pi}} (6\rho - \rho^2)e^{-\rho/3} \sin\theta \cos\varphi$$

$$(3py) = a^{-3/2} \frac{\sqrt{2}}{81\sqrt{\pi}} (6\rho - \rho^2)e^{-\rho/3} \sin\theta \sin\varphi$$

$$(3pz) = a^{-3/2} \frac{\sqrt{2}}{81\sqrt{\pi}} (6\rho - \rho^2) e^{-\rho/3} \cos\theta$$

$$(3d_{xz}) = a^{-3/2} \frac{\sqrt{2}}{81\sqrt{\pi}} \rho^2 e^{-\rho/3} \sin\theta \cos\theta \cos\varphi$$

$$(3d_{yz}) = a^{-3/2} \frac{\sqrt{2}}{81\sqrt{\pi}} \rho^2 e^{-\rho/3} \sin\theta \cos\theta \sin\varphi$$

$$(3d_{x^2-y^2}) = a^{-3/2} \frac{1}{81\sqrt{2\pi}} \rho^2 e^{-\rho/3} \sin^2\theta \cos 2\varphi$$

$$(3d_{xy}) = a^{-3/2} \frac{1}{81\sqrt{2\pi}} \rho^2 e^{-\rho/3} \sin^2\theta \sin 2\varphi$$

$$(3d_{z^2}) = a^{-3/2} \frac{1}{81\sqrt{6\pi}} \rho^2 e^{-\rho/3} (3\cos^2\theta - 1)$$

ANSWERS TO SELECTED PROBLEMS

1.1 (1) $\cos\theta = \sqrt{\frac{2}{5}} = 0.632$; $\theta = 50°46'$
(2) $\cos\theta = -\frac{1}{10}$; $\theta = 95°44'$

1.2 $\mathbf{E} = (0.447, -0.894, 0)$

1.3 (1) $\cos\theta_{ab} = \frac{1}{3}$; $\theta_{ab} = 70°33'$
(2) $\cos\theta_{ac} = -\frac{1}{3}$; $\theta_{ac} = 109°28'$

1.4 $AB = \begin{pmatrix} 0 & 0 \\ 0 & 0 \end{pmatrix}$; $BA = \begin{pmatrix} 1 & 1 \\ -1 & -1 \end{pmatrix}$

1.7 $\begin{pmatrix} 1 & 0 & 0 \\ 0 & 1 & 0 \\ 0 & 0 & 1 \end{pmatrix}$

1.9 $c_1 = 0$; $c_2 = -1$; $c_3 = \sqrt{5}$

1.10 (1) $\theta = 63°26'$; $|K\mathbf{v}|/|v| = \sqrt{5}$
(2) $\theta = 0$; $|K\mathbf{u}|/|u| = 3$

1.12 $-2, 1, 4$

1.13 $\lambda_1 = 5$; $\mathbf{E}_1 = \frac{1}{\sqrt{13}}(3, -2)$

$\lambda_2 = -8$; $\mathbf{E}_2 = \frac{1}{\sqrt{13}}(2, 3)$

1.14 $\lambda_1 = 3$; $\mathbf{E}_1 = \frac{1}{3}(1, 2, 2)$
$\lambda_2 = 6$; $\mathbf{E}_2 = \frac{1}{3}(2, 1, -2)$
$\lambda_3 = 9$; $\mathbf{E}_3 = \frac{1}{3}(2, -2, 1)$

1.19 $\lambda_1 = 9$; $\lambda_2 = \lambda_3 = -9$
$\mathbf{E}_1 = \frac{1}{3}(1, 2, -2)$

There are no unique solutions for $\mathbf{E}_2$, $\mathbf{E}_3$. Possible solutions are the pairs of vectors

$$\frac{1}{\sqrt{5}}(2, 0, 1); \quad \frac{1}{\sqrt{45}}(2, -5, -4)$$

$$\frac{1}{\sqrt{2}}(0, 1, 1); \quad \frac{1}{\sqrt{18}}(4, -1, 1)$$

1.20 (2) See Problem 1.14

2.1 (1) $E_1 = \alpha + \sqrt{2}\,\beta$; $\psi_1 = \frac{1}{2}(1, 2, 1)$

$E_2 = \alpha$; $\psi_2 = \frac{1}{\sqrt{2}}(1, 0, -1)$

$E_3 = \alpha - \sqrt{2}\,\beta$; $\psi_3 = \frac{1}{2}(1, -2, 1)$

(2) The bond energy of all species is $2\sqrt{2}\,\beta$.

2.2 (1) $E_1 = \alpha + 2\beta$; $E_2 = E_3 = \alpha$; $E_4 = \alpha - 2\beta$

$\psi_1 = \frac{1}{2}(1, 1, 1, 1)$; $\psi_4 = \frac{1}{2}(1, -1, 1, -1)$

For the degenerate level α there are no unique molecular orbitals, but possible solutions are the pairs

$$\tfrac{1}{2}(1, 1, -1, -1); \quad \tfrac{1}{2}(1, -1, -1, 1)$$

$$\frac{1}{\sqrt{2}}(1, 0, -1, 0); \quad \frac{1}{\sqrt{2}}(0, 1, 0, -1)$$

(2) Delocalization energy $= 0$. The strain energy in the small ring is thus uncompensated by any aromatic character in the molecule.

2.3 $p_{12} = p_{34} = 0.447$; $p_{23} = 0.722$

2.5 $\frac{2}{3}$

2.6 (1) $E_1 = \alpha + \frac{1}{2}(1 + \sqrt{5})\beta$; $\psi_1 = (0.526, 0.851)$

$E_2 = \alpha + \frac{1}{2}(1 - \sqrt{5})\beta$; $\psi_2 = (0.851, -0.526)$

(2) $\sqrt{5}\,\beta$

3.1 $x = 2(\sin x - \frac{1}{2}\sin 2x + \frac{1}{3}\sin 3x - \cdots)$

3.4 (1) $\langle p \rangle = 0$

(2) $\langle p^2 \rangle = \frac{1}{2}mh\nu$

3.6 $\delta x\, \delta p = \frac{1}{2}\hbar$

3.7 (1) $n \sim 4 \times 10^{11}$

(2) $n = 0$

4.1 (1) $\psi_{20} = (\pi x_0^2)^{-1/2}[1 - (r/x_0)^2] \exp[-\frac{1}{2}(r/x_0)^2]$

$\psi_{0\pm2} = (2\pi x_0^2)^{-1/2}(r/x_0)^2 \exp[-\frac{1}{2}(r/x_0)^2] \exp(2i\varphi)$

4.7 (1) $E_m = \frac{1}{2}\hbar^2 m^2/m_e r^2;\quad m = 0, \pm 1, \pm 2, \ldots$

$$\psi_m = \frac{1}{\sqrt{2\pi}} \exp(im\varphi)$$

5.6 (2) $\theta_{12} = \theta_{13} = \theta_{34} = 109°28'$. This is the tetrahedral bond angle exhibited by a set of sp^3 hybrid atomic orbitals. See Problem 1.3, part 2.

6.2 $E_1 = 2h\nu - \varepsilon\hbar;\quad \psi_1{}^0 = \frac{1}{\sqrt{2}}(\psi_x - i\psi_y)$

$E_2 = 2h\nu + \varepsilon\hbar;\quad \psi_2{}^0 = \frac{1}{\sqrt{2}}(\psi_x + i\psi_y)$

6.3 (1) $E_1 = 2h\nu + \frac{1}{2}\varepsilon x_0{}^2;\quad E_2 = 2h\nu + \frac{3}{2}\varepsilon x_0{}^2$
(2) $\psi_1{}^0 = \psi_x;\quad \psi_2{}^0 = \psi_y$

6.4 $E_2, E_2, E_2 \pm (e\mathscr{B}/2m_e c)\hbar$

8.2 (1) $\psi_0(1, 2) = \frac{1}{2\pi}\frac{1}{\sqrt{2}}[\alpha(1)\beta(2) - \alpha(2)\beta(1)]$

(2) $\psi_1(1, 2) = \frac{1}{2\pi}\frac{1}{\sqrt{2}}[\exp(i\varphi_2) - \exp(i\varphi_1)]\alpha(1)\alpha(2)$

$$\psi_2(1, 2) = \frac{1}{2\pi}\frac{1}{\sqrt{2}}[\exp(i\varphi_2)\alpha(1)\beta(2) - \exp(i\varphi_1)\alpha(2)\beta(1)]$$

$$\psi_3(1, 2) = \frac{1}{2\pi}\frac{1}{\sqrt{2}}[\exp(i\varphi_2)\beta(1)\alpha(2) - \exp(i\varphi_1)\beta(2)\alpha(1)]$$

$$\psi_4(1, 2) = \frac{1}{2\pi}\frac{1}{\sqrt{2}}[\exp(i\varphi_2) - \exp(i\varphi_1)]\beta(1)\beta(2)$$

There are four additional functions for the first excited state obtained from ψ_1 to ψ_4 by replacing $\exp(i\varphi)$ by $\exp(-i\varphi)$.

9.2 (1) $\lambda_1 = 1 + \sqrt{2};\quad \mathbf{E}_1 = \frac{1}{2}(1, \sqrt{2}, 1)$

$\lambda_2 = 1;\quad \mathbf{E}_2 = \frac{1}{\sqrt{2}}(1, 0, -1)$

$\lambda_3 = 1 - \sqrt{2};\quad \mathbf{E}_3 = \frac{1}{2}(1, -\sqrt{2}, 1)$

9.3 $\lambda_1 = 5;\quad \mathbf{E}_1 = \frac{1}{\sqrt{2}}(2, -1)$

$\lambda_2 = 1;\quad \mathbf{E}_2 = \frac{1}{\sqrt{2}}(0, 1)$

11.4 Monatomic

11.5 If the ordering of molecular energy states is as drawn in Figure 11.8 with $\pi 2p > \sigma 2p$, then C_2 should be paramagnetic. If $\sigma 2p > \pi 2p$, then C_2 should be diamagnetic. Experimentally C_2 is diamagnetic.

12.1 Symmetric class

12.2 There are three symmetry classes:

$$\left\{2s, 2px, \frac{1}{\sqrt{2}}[(1s)_a + (1s)_b]\right\}; \quad \left\{2py, \frac{1}{\sqrt{2}}[(1s)_a - (1s)_b]\right\}; \quad \{2pz\}$$

12.3 $a = 0.4475; \quad b = \pm 0.7071; \quad c = 0.5475; \quad d = 0; \quad n = 3.994$

12.4 $a = 0.7743; \quad b = 0; \quad c = -0.6328; \quad d = 0; \quad n = 0.668$

12.9 $\pi_1 = \frac{1}{2}(1, 1, 1, 1)$

$$\pi_2 = \frac{1}{\sqrt{2}}(0, 0, 1, -1)$$

$$\pi_3 = \frac{1}{\sqrt{6}}(0, 2, -1, -1)$$

$$\pi_4 = \frac{1}{2\sqrt{3}}(3, -1, -1, -1)$$

INDEX